Pathobiology

Founded 1938 as 'Schweizerische Zeitschrift für allgemeine Pathologie und Bakteriologie' by A. v. Albertini, A. Grumbach and H. Mooser, continued as 'Pathologia et Microbiologia' (1960–1975) and 'Experimental Cell Biology' (1976–1989); incorporating 'Pathology and Immunopathology Research', founded 1982 as 'Survey and Synthesis of Pathology Research' by J.M. Cruse and R.E. Lewis, continued as 'Pathobiology', edited by J.M. Cruse and R.E. Lewis (1990–1998)

Editor-in-Chief

Ch. Wittekind, Leipzig

Regional Editor: Far East

W. Yasui, Hiroshima

Editorial Board

B. Borisch, Geneva
A. Cardesa, Barcelona
H. Denk, Graz
W.E. Fleig, Halle
A. Katalinic, Lübeck
V. Keim, Leipzig
J.D. Kemp, Iowa City, Iowa
C.J. Kirkpatrick, Mainz
M.S. Lakshmi, Newcastle upon Tyne

W.J. Mergner, Baltimore, Md.
J. Neoptolemos, Liverpool
R. Pabst, Hannover
M. Reymond, Magdeburg
A. Tannapfel, Leipzig
M. Werner, Munich
T. Yoshiki, Sapporo
M. Zeitz, Homburg/Saar

ST. PHILIP'S COLLEGE LIBRARY

KARGER

Printed in Switzerland
on acid-free paper by
Reinhardt Druck, Basel

Appears bimonthly:
1 volume per year
(6 issues)

Guidelines for Authors

Pathobiology

Aims and Scope

Pathobiology offers a valuable forum for high quality original research into the pathophysiological and pathogenetic mechanisms underlying human disease. Aiming to serve as a bridge between basic biomedical research and clinical medicine, the journal welcomes articles from scientific areas such as pathology, oncology, anatomy, virology, internal medicine, surgery, cell and molecular biology, and immunology. Published bimonthly, the journal features original research papers, reviews, short communications and editorials. Papers highlighting the clinical relevance of pathological data are encouraged. A special section will be devoted to reports on methodological improvements, and in the 'International Forum' section current or controversial issues will be discussed by specialists in the field.

Submission

Manuscripts should be addressed to:
Prof. Dr. Ch. Wittekind
Institut für Pathologie
Universität Leipzig
Liebigstrasse 26
D–04103 Leipzig (Germany)

Manuscripts from the Far East to:
Prof. W. Yasui
Chairman, Department of Pathology
Hiroshima School of Medicine
1-2-3 Kasumi, Minami-ku
Hiroshima 734-8551 (Japan)

Manuscripts should be submitted in triplicate (with three sets of illustrations of which one is an original), typewritten double-spaced on one side of the paper, with a wide margin. The final version of your manuscript and illustration should also be submitted on disk. For details refer to 'Disk Submission' or www.karger.com/journals/disksub.htm. Editorial review and copy-editing will be executed from the disk version of your manuscript.

Conditions

Manuscripts must be written in English, and are all subject to peer review. Manuscripts are received with the explicit understanding that they are not submitted for simultaneous consideration by any other publication. Submission of an article for publication implies the transfer of the copyright from the author to the publisher upon acceptance. Accepted papers become the permanent property of 'Pathobiology' and may not be reproduced by any means, in whole or in part, without the written consent of the publisher. It is the author's responsibility to obtain permission to reproduce illustrations, tables, etc. from other publications. For papers that are accepted for publication, manuscript copies and illustrations will not be returned to the authors. For papers that are rejected, only the set of original figures will be returned.

Arrangement

Title page: The first page of each paper should indicate the title, the authors' names, and the institute where the work was conducted. A short title for use as running head as well as the full address of the author to whom correspondence should be sent are also required.

Full address: The exact postal address complete with postal code must be given at the bottom of the title page. Please also supply phone and fax numbers, as well as your e-mail address.

Key words: For indexing purposes, a list of 3–10 key words in English is essential

Abstract: **Each paper needs an abstract of up to 200 words structured with subheadings as follows: Objective(s), Methods, Results, Conclusion(s).**

Small type: Paragraphs which can or must be set in smaller type (case histories, test methods, etc.) should be indicated with a 'p' (petit) in the margin on the left-hand side.

Footnotes: Avoid footnotes. When essential, they are numbered consecutively and typed at the foot of the appropriate page.

Tables and illustrations: Tables and illustrations (both numbered in Arabic numerals) should be prepared on separate sheets. Tables require a heading and figures a legend, also prepared on a separate sheet. For the reproduction of illustrations, only good drawings and original photographs can be accepted; negatives or photocopies cannot be used. Due to technical reasons, figures with a screen background should not be submitted. When possible, group several illustrations on one block for reproduction (max. size 181×223 mm) or provide crop marks. On the back of each illustration, indicate its number, the author's name, and 'top' with a soft pencil.

Color illustrations: Up to 6 color illustrations per page can be integrated within the text at the reduced price of CHF 660.– / USD 545.00 per page. Color illustrations are reproduced at the author's expense.

References: In the text identify references by Arabic numerals [in square brackets]. Material submitted for publication but not yet accepted should be noted as 'unpublished data' and not be included in the reference list. The list of references should include only those publications which are cited in the text. Do not alphabetize; number references in the order in which they are first mentioned in the text. The surnames of the authors followed by initials should be given. There should be no punctuation other than a comma to separate the authors. Cite all authors, 'et al' is not sufficient. Abbreviate journal names according to the Index Medicus system. (Also see International Committee of Medical Journal Editors: Uniform requirements for manuscripts submitted to biomedical journals. N Engl J Med 1997;336:309–315.)

Examples
(a) *Papers published in periodicals:* Sun J, Koto H, Chung KF: Interaction of ozone and allergen challenges on bronchial responsiveness and inflammation in sensitised guinea pigs. Int Arch Allergy Immunol 1997;112:191–195.
(b) *Monographs:* Matthews DE, Farewell VT: Using and Understanding Medical Statistics, ed 3, revised. Basel, Karger, 1996.
(c) *Edited books:* Parren PWHI, Burton DR: Antibodies against HIV-1 from phage display libraries: Mapping of an immune response and progress towards antiviral immunotherapy; in Capra JD (ed): Antibody Engineering. Chem Immunol. Basel, Karger, 1997, vol 65, pp 18–56.
(d) *Papers published in electronic format:* Black CA: Delayed type hypersensitivity: Current theories with a historic perspective. Dermatol Online J 1999;5:7. http://dermatology.cdlib.org/DOJvol5num1.

Page charges

Page charges may be levied if the allotted size of the manuscript, as stipulated by the editors of the journal, is exceeded. Each additional complete or partial page is charged to the author at CHF 275.– / USD 196.00. 3 manuscript pages (including tables, illustrations and references) are equal to approximately one printed page.

Galley proofs

Unless indicated otherwise, galley proofs are sent to the first-named author and should be returned with the least possible delay. Alterations made in galley proofs, other than the correction of printer's errors, are charged to the author. No page proofs are supplied.

Reprints

Order forms and a price list are sent with the galley proofs. Orders submitted after the issue is printed are subject to considerably higher prices.

KARGER

© 2001 S. Karger AG, Basel

Fax + 41 61 306 12 34
E-Mail karger@karger.ch
www.karger.com

The Guidelines for Authors are available at:
www.karger.com/journals/pat/pat_gl.htm

Pathobiology

General Information

Publication data: Volume 68, 2000 of 'Pathobiology' appears with 6 issues.

Subscription rates: Subscriptions run for a full calendar year. Prices are given per volume.
Personal subscription:
Subscription + postage and handling
CHF 240.– + CHF 25.90 Europe
CHF 240.– + CHF 39.10 Overseas
USD 183.50 + USD 25.90
+ online subscription per volume:
CHF 39.–/USD 30.00
(Must be in the name of, billed to, and paid by an individual. Order must be marked 'personal subscription'.)
Institutional subscription:
Subscription + postage and handling
CHF 960.– + CHF 32.30 Europe
CHF 960.– + CHF 48.80 Overseas
USD 734.00 + USD 32.30
+ online subscription per volume:
CHF 58.–/USD 45.00
Airmail postage: CHF 36.–, USD 30.00.
(Extra per volume.)
Discount subscription prices:
Members of the AAI.

Microforms: Please direct all enquiries and orders to: University Microfilms Inc.
300 North Zeeb Road, P.O. Box 1346
Ann Arbor, MI 48106-1346 (USA)

Subscription orders: Orders can be placed at agencies, bookstores, directly with the Publisher
S. Karger AG, P.O. Box,
CH–4009 Basel (Switzerland)
or with any of the following distributors:
Deutschland: S. Karger GmbH, Postfach
D–79095 Freiburg
France: Librairie Luginbühl
36, bd de Latour-Maubourg, F–75007 Paris
Great Britain, Ireland: S. Karger Publishers
58 Grove Hill Road
Tunbridge Wells, Kent TN1 1SP, UK
USA: S. Karger Publishers, Inc.
26 West Avon Road, P.O. Box 529
Farmington, CT 06085, USA
India, Bangladesh, Sri Lanka:
Panther Publishers Private Ltd.
39, 6th Cross
Wilson Gardens
Bangalore 560 027, India
Japan: Katakura Libri, Inc.
2F, Marushima Bldg, 4-6-17, Yushima,
Bunkyo-ku, Tokyo 113-0034, Japan

Change of address: Both old and new address should be sent to the subscription source.

Single issues and back volumes: Information can be obtained through the Publisher.

Advertising: Correspondence and rate requests should be addressed to the Publisher.

Bibliographic indices: This journal is regularly listed in bibliographic services, including *Current Contents®*.

Copying: This journal has been registered with the Copyright Clearance Center (CCC), as indicated by the code appearing on the first page of each article.
For readers in the US, this code signals consent for copying of articles for personal or internal use, or for the personal or internal use of specific clients, provided that the stated fee is paid per copy directly to
Copyright Clearance Center Inc.
222 Rosewood Drive
Danvers, MA 01923 (USA)
A copy of the first page of the article must accompany payment. Consent does not extend to copying for general distribution, for promotion, for creating new works, or for resale. In these cases, specific written permission must be obtained from the copyright owner,
S. Karger AG,
P.O. Box, CH–4009 Basel (Switzerland).

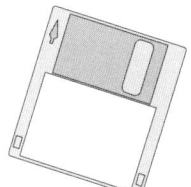

Disk Submission

A **comprehensive version** of these guidelines for the **submission of manuscripts and illustrations on disk** can be accessed online under www.karger.com/journals/disksub.htm or requested as a leaflet from the editorial department.

Please follow the general rules on style and arrangement in the **'Guidelines for Authors'** of the journal. Submit a **double-spaced printout** of your manuscript matching the file on disk exactly.

Technical Information
Store only data pertinent to this manuscript. Include your fonts (screen and printer fonts for Macintosh).
For a **text only** manuscript, use a 3.5-inch disk.
For **text and illustrations,** use 3.5 or 5.25 Magneto Optical Disc; CD-ROM (DOS or hybrid); iomage Zip/DOS (100 MB), or 44 MB, 88 MB or 200 MB SyQuest. Compress extensive data using ZIP format.

Data Description
Label your disk with the **title of the journal or book** for which it is submitted; a **short version** of the paper's title; your **name,** postal and e-mail address, fax and phone number; **hard and software** used, including the version; and the **complete list of the file names.** File names consist of eight characters before the dot identifying the file and three characters after the dot identifying the application.

Text and Format
The **preferred word-processing package is MS-Word.** Other commonly used PC text programs and ASCII formats are accepted.
Update your data when you revise subsequent to the reviewer's or editor's decision and mark revisions clearly. Manuscripts accepted for publication are subject to **copy editing:** changes and/or questions will be marked in the manuscript, which will be returned to you with your disk and the galley proofs for checking.
Enter your text continuously flush left. Do not use any layout functions. Do not split words at the end of a line. **Do not use indentations.** Use a blank line before and a hard return at the end of paragraphs.
Arrange headings flush left. Insert only **one space after punctuation marks.** Insert Greek letters, mathematical symbols etc. with your word-processing program. Use **boldface and italics** as well as sub- and superscript where appropriate.
Tables are part of the text and should be placed at the end of the text file. Separate the individual parts of a table by lines. Use the tabulator instead of spaces. Format operation data on one line in the same column. **Legends to illustrations are part of the text** and should also be listed at the end of the text file.

Illustrations
Store illustrations as separate files: do not integrate them in your text. Export **line drawings/vector graphics** as TIF, EPS or WMF format. Use Photoshop for processing and retouching **scanned halftone images.** Save the original scan and the processed version. Export **black and white or color images** as TIF or EPS format in their anticipated size in print.
For detailed information on how to submit illustration data on disk please consult the leaflet or www.karger.com/journals/disksub.htm.

KARGER

© 2001 S. Karger AG, Basel

Fax + 41 61 306 12 34
E-Mail karger@karger.ch
www.karger.com

The Journal Home Page is available at:
www.karger.com/journals/pat

A color atlas based on more than 1,600 aspiration cytology biopsies

Salivary Gland Tumours

Jerzy Klijanienko
Philippe Vielh

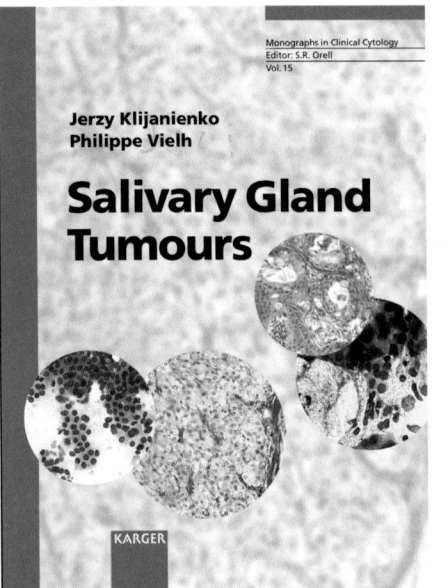

This monograph is based on a detailed review of more than 1,600 aspiration cytology biopsies of salivary gland tumors, including even rare entities, which were collected over a period of 45 years at the Institut Curie and correlated with histopathological diagnoses according to the most recent classifications (WHO and AFIP). Review of such a large series of cases has allowed the authors to specify diagnostic criteria for the great variety of tumor and tumor-like entities occurring at this site, including the rare ones. In order to facilitate cytological diagnosis, lesions are divided into adenomas, low-grade and high-grade malignancies, other tumors and tumor-like lesions, and further subclassified according to the dominant cell types. Both typical and unusual features are described, illustrated, and summarized in tables and key points for differential diagnosis. Guidelines for the practical clinical application of aspiration cytology related to salivary gland tumors as well as the accuracy of cytological diagnosis and a complete bibliography are presented. Preoperative cytological diagnosis may reduce the number of unnecessary surgical removals and avoid frozen sections which are notoriously difficult.

Illustrated with numerous color photographs and representing the most up-to-date text of its kind, this handbook is a valuable resource primarily for surgical pathologists in the ENT field, but will also be of interest to surgeons, particularly those involved in head and neck pathology, and medical oncologists.

Main Headings

1. Introduction
Demographics of Salivary Gland Tumours
History and Current Aspects of Fine-Needle Aspiration Cytology of Salivary Tumours
Accuracy of Fine-Needle Aspiration Cytology in Salivary Tumours
Clinical Indications for Fine-Needle Aspiration Cytology
Complications of Fine-Needle Aspiration Cytology
2. Fine Needle Aspiration Cytology Procedure and Ancillary Techniques
Specimen Processing
Choice of Ancillary Techniques
3. Imaging of Salivary Gland Tumours
4. Salivary Gland Anatomy and Tumour Histogenesis
Structure and Anatomy
Tumour Histogenesis
5. Basic Cytological Components and Diagnostic Approach
Components
Tumour Classification
Diagnostic Approach
6. Adenomas
Adenomas with Prominent Myoepithelial Cells
Adenomas with Prominent Oncocytes
Adenomas with Prominent Basaloid Cells (Cell and Canalicular Adenomas)
Adenomas with Prominent Sebaceous Cells
Papillary and Cystic Adenomas

7. High- and Intermediate-Grade Carcinomas
Carcinomas with Squamous Cells
Undifferentiated Carcinoma
Carcinomas with Prominent Myoepithelial Cells
Carcinomas with Oncocytes
Carcinomas with Sebaceous Cells
Other Carcinomas
8. Low-Grade Carcinomas
Low-Grade Mucoepidermoid Carcinoma
Acinic Cell Carcinoma
Polymorphous Low-Grade Adenocarcinoma
Papillary Cystadenocarcinoma
Basal Cell Adenocarcinoma
9. Benign and Malignant Mesenchymal Tumours and Miscellaneous Lesions
Benign Mesenchymal Tumours
Malignant Mesenchymal Tumours
Miscellaneous Lesions
10. Lymphomas
11. Secondary Tumours
12. Tumour-Like Lesions
Sialadenosis
Necrotizing Sialometaplasia
Lymphoepithelial Lesions
Salivary Gland Cysts
Chronic Sclerosing Sialadenitis (Küttner's Tumour)
Reactive Processes

www.karger.com/bookseries/moclc

Monographs in Clinical Cytology, Vol. 15
Series Editor: Orell, S.R. (Kent Town)
ISSN 0077–0809

Klijanienko, J.; Vielh P. (Paris)
Salivary Gland Tumours
XII + 142 p., 215 fig., 212 in color, 52 tab., hard cover, 2000
CHF 198.– / DEM 257.– / USD 172.25
Prices subject to change
DEM price for Germany, USD price for USA only
ISBN 3–8055–7024–4

Order Form

➤ Please send: ____ copy/ies
Postage and handling free with prepayment

Payment:
☐ Check enclosed   ☐ Please bill me

Please charge to my credit card
☐ American Express   ☐ Diners   ☐ Eurocard
☐ MasterCard   ☐ Visa

Card No.: _____

Exp. date: _____

Orders may be placed with any bookshop, subscription agency, directly with the publisher or through a Karger distributor.

Fax to: +41 61 306 12 34

S. Karger AG, P.O. Box, CH–4009 Basel (Switzerland)
E-Mail orders@karger.ch, **www.karger.com**

Name/Address:
_____
_____
_____
_____

Date: _____

Signature: _____

Pathobiology

Laser Microdissection

2nd International Conference on Laser Microdissection
Geneva, September 21–23, 2000

Editor
Bettina Borisch, Geneva

33 figures, 6 in color, and 6 tables, 2001

Basel · Freiburg · Paris · London · New York ·
New Delhi · Bangkok · Singapore · Tokyo · Sydney

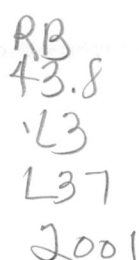

S. Karger
Medical and Scientific Publishers
Basel · Freiburg · Paris · London
New York · New Delhi · Bangkok
Singapore · Tokyo · Sydney

Drug Dosage
The authors and the publisher have exerted every effort to ensure that drug selection and dosage set forth in this text are in accord with current recommendations and practice at the time of publication. However, in view of ongoing research, changes in government regulations, and the constant flow of information relating to drug therapy and drug reactions, the reader is urged to check the package insert for each drug for any change in indications and dosage and for added warnings and precautions. This is particularly important when the recommended agent is a new and/or infrequently employed drug.

All rights reserved.
No part of this publication may be translated into other languages, reproduced or utilized in any form or by any means, electronic or mechanical, including photocopying, recording, microcopying, or by any information storage and retrieval system, without permission in writing from the publisher or, in the case of photocopying, direct payment of a specified fee to the Copyright Clearance Center (see 'General Information').

© Copyright 2001 by S. Karger AG,
P.O. Box, CH–4009 Basel (Switzerland)
Printed in Switzerland on acid-free paper by
Reinhardt Druck, Basel
ISBN 3–8055–7208–5

KARGER

Fax + 41 61 306 12 34
E-Mail karger@karger.ch
www.karger.com

Contents

163 Introduction
Borisch, B. (Geneva)

165 Oligoclonality of Early Lesions of the Urothelium as Determined by Microdissection-Supported Genetic Analysis
Stoehr, R.; Hartmann, A.; Hiendlmeyer, E.; Mürle, K.; Wieland, W.; Knuechel, R. (Regensburg)

173 Detection of Gene Amplification in Intraductal and Infiltrating Breast Cancer by Laser-Assisted Microdissection and Quantitative Real-Time PCR
Glöckner, S.; Lehmann, U.; Wilke, N.; Kleeberger, W.; Länger, F.; Kreipe, H. (Hannover)

180 Laser Microdissection and Microsatellite Analyses of Breast Cancer Reveal a High Degree of Tumor Heterogeneity
Wild, P.; Knuechel, R.; Dietmaier, W.; Hofstaedter, F.; Hartmann, A. (Regensburg)

191 Cell Type-Specific mRNA Quantitation in Non-Neoplastic Tissues after Laser-Assisted Cell Picking
Bohle, R.M.; Hartmann, E.; Kinfe, T.; Ermert, L.; Seeger, W.; Fink, L. (Giessen)

196 Laser-Assisted Microdissection and Short Tandem Repeat PCR for the Investigation of Graft Chimerism after Solid Organ Transplantation
Kleeberger, W.; Rothämel, T.; Glöckner, S.; Lehmann, U.; Kreipe, H. (Hannover)

202 Quantitative Molecular Analysis of Laser-Microdissected Paraffin-Embedded Human Tissues
Lehmann, U.; Bock, O.; Glöckner, S.; Kreipe, H. (Hannover)

209 Laser Capture Microdissection: Methodical Aspects and Applications with Emphasis on Immuno-Laser Capture Microdissection
Fend, F. (Munich); Kremer, M. (Munich/Oberschleissheim); Quintanilla-Martinez, L. (Oberschleissheim)

215 Recovering DNA and Optimizing PCR Conditions from Microdissected Formalin-Fixed and Paraffin-Embedded Materials
Ren, Z.-P.; Sällström, J.; Sundström, C.; Nistér, M.; Olsson, Y. (Uppsala)

218 Diagnosis of Papillary Thyroid Carcinoma Is Facilitated by Using an RT-PCR Approach on Laser-Microdissected Archival Material to Detect RET Oncogene Activation
Lahr, G.; Stich, M.; Schütze, K.; Blümel, P.; Pösl, H.; Nathrath, W.B.J. (Munich)

227 Microsatellite Instability in Tumor and Nonneoplastic Colorectal Cells from Hereditary Non-Polyposis Colorectal Cancer and Sporadic High Microsatellite-Instable Tumor Patients
Dietmaier, W.; Gänsbauer, S. (Regensburg); Beyser, K. (Kassel); Renke, B.; Hartmann, A.; Rümmele, P.; Jauch, K.-W.; Hofstädter, F. (Regensburg); Rüschoff, J. (Kassel)

232 Laser Microdissection as a New Approach to Prefertilization Genetic Diagnosis
Clement-Sengewald, A.; Buchholz, T. (Munich); Schütze, K. (Bernried)

237 Author Index Vol. 68, No. 4–5, 2000
238 Subject Index Vol. 68, No. 4–5, 2000

Introduction

Bettina Borisch

The 2nd International Conference on Laser Microdissection took place in Chavannes-de-Bogis near Geneva. Since the first meeting held in Les Diablerets in April 1998, much progress has been made both in the laser microdissection microscope technology as well as in the downstream methods of analyzing the microdissected material.

In addition to a strong European participation, about 80 participants from 14 countries including Asia, Australia and America attended the meeting. The opportunities offered by the workshops were highly appreciated and allowed every participant to have hands-on experience with four different laser microdissection microscopes.

The present volume contains original articles that cover the contents of talks given at the meeting. The reader will no doubt appreciate the broad variety of topics discussing the powerful tool of microdissection. It will be readily understood that the development of micromethods in analytic and diagnostic biomedicine (such as chips/arrays, MALDI) requires morphologically verified and analyzed 'micromaterial'. Almost all questions addressed to whole tissue preparations some years ago are no longer useful and have to be redirected to well-defined parts of tissues or structures.

The first session on solid tumours readily underlined this aspect among other things. The opening presentation by Lemoine [1] gave an overview of the opportunities now available for the study of solid tumours and illustrated the (inherited) pancreatic carcinoma and the hepatocellular carcinoma. Studying precursor lesions such as regenerative nodules of the liver and intraductal pancreatic mucinous tumours using chip technology, he coined the term of 'expressome' for this way of addressing the tumour biology of particular situations as a model for investigating solid tumours in general. The following presentations in the session were concerned with the problems of clonality and microheterogeneity of solid tumours. Knüchel et al. (p. 165) discussed the importance of the clonality during the establishment of urothelial cancerous lesions. An important part of the meeting was dedicated to breast carcinoma and its precursor lesions. Glöckner et al. (p. 173) described the combined use of microdissection and real-time PCR to study intraductal lesions and their adjacent invasive components. Diallo [2] showed that the HUMARA assay for clonality in breast lesions might be of limited value as terminal ductal lobular units are monoclonal to start with. These findings indeed support the hypothesis of distinct stem cell-derived monoclonal patches as a model for the organization of the normal mammary gland. By analyzing microsatellite instability (MSI) and loss of heterozygosity (LOH), Hartmann et al. (p. 180) showed a high degree of heterogeneity in breast tumours.

The haematopathology session was introduced by Hansmann [3, 4] who was a pioneer in microdissection and single cell analysis, starting out with manual microdissection of Hodgkin cells. Besides the old data on HD/RS cells and their origin, new results on composite lymphomas and lymphocyte-predominant Hodgkin's disease as well as the T cell aspect of Hodgkin's disease were presented. The second presentation (Fey et al.) summarized the results obtained on LOH of HD and RS cells. Microsatellites at 1q42, 4q26, 9p23 and 11q23 were analyzed. Clonal LOH was seen in all but one case, and the locus most frequently altered was 4q26 [5]. Finally Leoncini et al. [6] presented the case of a patient with simultaneously occurring HD and large B cell lymphoma. Their analysis may suggest that RS and large B cell lymphoma cells originate from a common precursor in which a secondary V_H replacement took place during germinal centre reaction.

The following session showed that microanalysis is not only useful in tumour cell biology but also in the analysis of non-neoplastic tissues. After an overview talk by Kölble [7], Bohle et al. (p. 191) presented an animal model which allows not only a cell type-specific mRNA quantification but also a cell type-specific characterization by combining this analysis with immunostaining. The Hannover group (Kleeberger et al., p. 196) showed that using a polymorphic short tandem repeat marker (STR) which is widely used in forensic medicine (SE33), graft chimerism after solid organ transplantation can be evaluated.

The keynote lecture by Liotta [8] covered the developments to come: functional proteomics, proteomics applied to molecular targets and protein pathway profiling. Combining these tools should allow us to describe more closely the functional state of a given protein pathway. He foresees that these questions will be answered using microdissection and combined downstream tools while concentrating especially on the proteomic approach. Lehmann and Kreipe (p. 202) pointed out that while doing a quantitative molecular analysis of laser-microdissected human tissues epigenetic modifications such as methylation can be evaluated. Fend (p. 209) gave a thorough review of the combination of immunohistochemistry and laser capture microdissection with all its pros and cons. More on the technical aspects, Ren et al. (p. 215) showed how they optimized the PCR conditions for microdissected formalin-fixed and paraffin-embedded materials.

In some instances, these new techniques are already used for diagnostics. Lahr et al. (p. 218) showed the usefulness of laser microdissection in the diagnosis of papillary thyroid carcinoma using an RT-PCR approach in archival material, and Rubbia et al. [9] the microanalysis of hepatic lesions. Dietmaier et al. (p. 227) discussed the application of LOH and MSI on microdissected colorectal tumours. Crnogorac-Jurcevic [1] extended the data of the Lemoine group by presenting cDNA arrays for generating a comprehensive gene expression profile of pancreatic adenocarcinoma. An important contribution (Clement-Sengewald et al., p. 232) focused on another application of laser microdissection, namely the use of the laser to extract polar bodies for a prefertilization diagnosis.

The microaspects being extensively covered, the macroaspects of life were highlighted by the sunny September weather that showed magnificent views of the Mont-Blanc and offered excellent opportunities to do tours to nearby vineyards and restaurants. The articles published herein give a complete overview of the present state of the art in laser microdissection and its applications and developments.

References

1 Sirivatanauskorn Y, Drury R, Crnogorac-Jurcevic T, Sirivatanauskorn V, Lemoine NR: Laser assisted-microdissection: Applications in molecular pathology. J Pathol 1999;1892:150–154.

2 Diallo R, Schaefer K-L, Poremba C, Shivazi N, Willmann V, Buerger H, Dockhorn-Dworniczak B, Boecker W: Monoclonality in the normal epithelium and in hyperplastic and neoplastic lesions of the breast. J Pathol 2001;193: 27–32.

3 Küppers R, Rajewsky K, Zhao M, Simons G, Laumann R, Fischer R, Hansmann M-L: Hodgkin disease: Hodgkin- and Reed-Sternberg cells picked from histological sections show clonal immunoglobulin rearrangements and appear to be derived from B cells at various stages of development. Proc Natl Acad Sci USA 1994;91:10962–10966.

4 Küppers R, Klein U, Hansmann M-L, Rajewsky K: Cellular origin of human B-cell lymphomas. N Engl J Med 1999;341:1520–1529.

5 Hasse U, Tinguely M, Leibundgut EO, Cajot JF, Garvin AM, Tobler A, Borisch B, Fey MF: Clonal loss of heterozygosity in microdissected Hodgkin and Reed-Sternberg cells. J Natl Cancer Inst 1999;91:1581–1583.

6 Bellan C, Lazzi S, Zazzi M, Lalinga AV, Palummo N, Galieni P, Marafioti T, Tonini T, Cinti C, Leoncini L, Pileri SA, Tosi P: Immunoglobulin gene rearrangement analysis in composite Hodgkin's disease and large B-cell-lymphoma. Evidence for receptor revision of immunoglobulin heavy chain variable region genes in Hodgkin-Reed-Sternberg cells ? Diagn Mol Pathol, in press.

7 Kölble K: The LEICA microdissection system: Design and applications. J Mol Med 2000;787: B24–B25.

8 Emmert-Buck MR, Strausberg RL, Krizman DB, Bonaldo MF, Bonner RF, Bostwick DG, Brown MR, Buetow KH, Chuaqui RF, Cole KA, Duray PH, Englert CR, Gillespie JW, Greenhut S, Grouse L, Hillier LW, Katz KS, Klausner RD, Kuznetzov V, Lash AE, Lennon G, Linehan WM, Liotta LA, Marra MA, Munson PJ, Ornstein DK, Prabhu VV, Prange C, Schuler GD, Soares MB, Tolstoshev CM, Vocke CD, Waterston RH: Molecular profiling of clinical tissue specimens: Feasibility and applications. Am J Pathol 2000;156:1109–1115.

9 Paradis V, Dargere D, Bonvoust F, Rubbia-Brandt L, Ba N, Bioulac-Sage P, Bedossa P: Clonal analysis of micronodules in virus C-induced liver cirrhosis using laser capture microdissection (LCM) and HUMARA assay. Lab Invest 2000;80:1553–1559.

Oligoclonality of Early Lesions of the Urothelium as Determined by Microdissection-Supported Genetic Analysis

Robert Stoehr[a] Arndt Hartmann[a] Elke Hiendlmeyer[a] Kristin Mürle[a]
Wolfgang Wieland[b] Ruth Knuechel[a]

[a]Institute of Pathology, University of Regensburg, and [b]Department of Urology, St. Josef Hospital, Regensburg, Germany

Key Words

Urothelial lesions · Microdissection · Fluorescence in situ hybridization · Loss of heterozygosity · Clonality

Abstract

Aim: To contribute to the ongoing discussion of clonality of human urothelial cancer it was considered a valuable approach to analyze multiple areas from cystectomy specimens for deletions of chromosomes known to be involved early in bladder cancer development. *Material and Methods:* Thus, in 86 biopsies of 4 human cystectomies with different histological findings (maximal diagnosis: pT1G2, pTaG3, pT2G2, normal) loss of heterozygosity (LOH) was investigated as a deletion marker using markers of chromosomes 8p, 9p, 9q and 17p. Findings were compared to histology of the lesion. *Results:* Findings indicate: (1) no changes in the markers investigated in the bladder with histologically normal urothelium in contrast to detection of LOH in normal urothelium of tumour-bearing bladders; (2) an accumulation of the number of LOH with increasing malignancy of lesions within one bladder, and (3) indications of oligoclonal neoplastic lesions in two of the urinary bladders investigated. *Conclusions:* The investigation of multiple lesions within one bladder presents a snapshot of genetic changes in differently advanced tumour stages. The hypotheses of tumour evolution and oligoclonality as derived from our LOH data need to be supported by deletion-independent clonality studies as X-chromosomal inactivation analysis.

Copyright © 2001 S. Karger AG, Basel

Introduction

The clinically determining features of urothelial cancer are multifocality and frequent recurrence. Multifocal tumours often present with varying degrees of stage and grade, and are also known to be frequently accompanied by other urothelial lesions such as hyperplasias, dysplasias and carcinomata in situ. The causal relationship between the lesions is a matter of ongoing discussion. Partially there is evidence of a clonal origin of these lesions [1], which is in contrast to a multifocally arising tumour presenting with several clones [2]. Increasing numbers of genetic analyses of tumours of a different stage and grade have led to the hypothesis that papillary tumours may arise through different initial genetic events than carcinomata in situ with consequent invasive solid tumours [3].

While most data are derived from manifest or advanced tumour stages it seems necessary to also look at early bladder lesions and normal urothelium to support the hypotheses on the initiation of bladder cancer. Fluorescence diagnosis of bladder lesions with 5-aminolevulinic acid has helped recently to identify higher frequencies of early bladder lesions, e.g. increasing the overall number of carcinomata in situ by 30% [4, 5]. Analysis of fluorescence-guided biopsies with the diagnosis of hyperplasia in comparison with synchronous and consequent non-muscle invasive papillary tumours revealed genetic changes less or identical to those of the tumours in 75% of cases [6]. These investigations were carried out by multicolour fluorescence in situ hybridization (FISH). Another study using FISH and deletion mapping of the same regions in multifocal early papillary tumours, also based on fluorescence-guided endoscopy, indicated two clones in one patient, while the majority of patients showed only one tumour clone [7]. From these observations it seems reasonable to assume that a more thorough mapping of whole urinary bladder specimens obtained from tumour cystectomies would give additional information, especially since a larger spectrum of lesions will be obtained for investigation. This has recently been applied by Czerniak et al. [8], who documented the pattern of chromosome 9 deletions by using a set of primers for 52 microsatellite loci for both chromosome arms in the bladder urothelium and compared it to bladder washing and urine cytology specimens.

The study presented here shows data on 4 cystectomy specimens, in which an overall of 86 biopsies were analyzed. Since normal urothelium and dysplastic lesions were included, the DNA of small amounts of microdissected tissue was amplified reproducibly by whole genome amplification using an improved primer extension preamplification PCR (I-PEP-PCR) [9]. This is considered an important support for this type of analysis enabling tests of a number of microsatellite markers for loci of chromosomes 8, 9 and 17 as shown here, or for the extension of the analysis to other techniques such as comparative genomic hybridization or sequencing from the same source DNA.

Material and Methods

Specimens and Histopathology

Cystectomy specimens were obtained immediately from the survey room (Department of Urology, St. Josef Hospital, Regensburg), and cut open with an inverted Y section by an experienced histopathologist. Fresh samples were obtained in a clockwise circular fashion from macroscopically normal bladder lesions, followed by systematic sampling of macroscopically suspicious areas and overt tumour. While these samples were immediately snap-frozen, adjacent tissue of each block was formalin-fixed, and processed for routine histopathology diagnosis. The individual diagnosis was based on routine samples of paraffin-embedded material and reference sections for the frozen tissue biopsies, both stained with haematoxylin-eosin. Staging was performed according to the International Union against Cancer (UICC) [10] and grading according to the World Health Organization [11], since diagnoses were established before the recent release of the WHO classification [12]. In addition, the histopathology of the frozen sections together with its location was documented as a basis for comparison of the genetic analysis of this lesion.

Isolation of Urothelial Cells and DNA

Serial sections (5 µm) of frozen biopsies adjacent to the reference sections were cut, stained with methylene blue for 15 s, and urothelial cells were dissected from stromal cells. This dissection was carried out either manually with a needle under an inverted microscope, or in case of small sample volume, or delicate invasive tumour cell strands by laser microdissection [PALM, Wolfrathshausen, Germany; see 13]. For PCR analysis pure epithelial populations and normal non-urothelial tissue (from vagina or perivesical tissue) were processed immediately as described before [7].

Whole Genome Amplification

The protocol for an improved primer extension preamplification technique was developed by Dietmaier et al. [9] using an MJR PTC200 thermocycler (Biozym, Oldenburg, Germany). In brief, after proteinase K digestion 50 amplification cycles were performed, each consisting of a 1-min step at 94°C, a 2-min step at 37°C, a ramping step of 0.1°C/s to 55°C, a 4-min step at 55°C, and a 30-second step at 68°C. I-PEP-PCR was set up by adding 50 µl I-PEP mix (final concentration 0.05 mg/ml gelatine, 16 µmol/l totally degenerate 15-nucleotide-long primer; MWG Biotech, Ebersberg, Germany; 0.1 mmol/l dNTP, 3.6 U Expand High Fidelity polymerase, Boehringer, Mannheim, Germany; 2.5 mmol/l MgCl$_2$ in 1 × PCR buffer No. 0.3, provided with the polymerase) to 10 µl lysed tissue.

Microsatellite Analysis

Specific microsatellite PCR (0.2 mmol/l dNTP, 0.3 µmol/l primers, 0.5 U Taq polymerase, Life Technologies, 1.5 mmol/l MgCl$_2$) was performed using 2 µl of the preamplified PCR as template in a final volume of 20 or 30 µl, respectively, in a PTC100 thermocycler (MJ Research, Watertown, Md.) for 35 cycles: 94°C for 1 min, 50–60°C for 1 min, 72°C for 1 min, followed by a final extension at 72°C for 8 min. Amplified microsatellites were detected by polyacrylamide gel electrophoresis and silver staining as described [14]. The silver-stained gels were assessed visually, and informative cases were scored as allelic loss when intensity of the signal for a tumour allele was decreased to 50% relative to the matched normal allele. Microsatellite instability was defined as the occurrence of additional alleles in the tumour tissue compared with the normal DNA. Thirteen microsatellites were used for loss of heterozygosity (LOH) analyses. Primers were obtained from MWG Biotech. Primer sequences and annealing temperatures were as follows: p53alu (17p13.1): 5′AGGAGGTTGCAGTAAGCGGA3′ and 5′AACAGCTCCTTTAATGGCAG3′ at 60°C [15], D9S304 (9p21): 5′GTGCACCTCTACACCCAGAC3′ and 5′TGTGCCCACACACATCTATC3′ at 60°C, D9S1751 (Pky11, 9p21): 5′TTGTTGATTCTGCCTTCAAAGTCTTTTAAC3′ and 5′CGTTAAGTCCTCTATTACACAGAG3′ at 55°C, D9S1747 (Pky2, 9p21): 5′ATTCAACGAGTGGGATGAAG3′ and 5′TCCAGGTTGCTGCAAATGCC3′ at 55°C, D9S1748 (Pky3, 9p21): 5′CACCTCAGAAGTCAGTGAGT3′ and 5′GTGCTTGAAATACACCTTTCC3′ at 55°C, D9S1752 (Pky15, 9p21): 5′AGACTACACAGGATGAGGTG3′ and 5′GCAAGTCATAAGGGGATTTC3′ at 60°C [16], D9S171 (9p21): 5′AGCTAAGTGA-

ACCTCATCTCTGTCT3′ and 5′ACCCTAGCACTGATGGTA-TAGTCT3′ at 55°C, D9S303 (9q21): 5′CAACAAAGCAAGA-TCCCTTC3′ and 5′TAGGTACTTGGAAACTCTTGGC3′ at 55°C, D9S747 (9q32): 5′GCCATTATTGACTCTGGAAAAGAC3′ and 5′CAGGCTCTCAAAATATGAACAAAT3′ at 56°C, D9S905 (9q34.1): 5′GTGGGAAAATTGGCCTAAGT3′ and 5′CTTCT-GAGCCTCACACCTGT3′ at 63°C, D9S65 (9q34.2): 5′CCTT-GCAGACTGATGGAGAA3′ and 5′GCGGACAATTAGGTTT-CAGG3′ at 56°C, D8S1106 (8p22): 5′TTGTTTACCCCTG-CATCACT3′ and 5′TTCTCAGAATTGCTCATAGTGC3′ at 56°C, D8S311 (8p12): 5′GCTGAAGGCAAGAGAATCGCT3′ and 5′TGCTCTT-GGGGATGTTGGTGAAATC3′ at 58°C. To avoid errors due to preferential amplification of one allele during PCR, all LOH analyses were run in duplicates following independent PCR reactions for the DNA aliquots.

Results

Overall, 86 samples covering representative areas of the entire urothelial lining of 4 cystectomy specimens from female patients with infiltrative bladder cancer were investigated. For LOH analyses samples were used when they showed heterozygosity in at least one microsatellite marker on 9p and 9q, 17p and 8p, and reproducible PCR amplification results could be obtained. The distribution of the informative samples in LOH analysis is shown in table 1. The number of the analyzed samples for each chromosome varied depending on the number of heterozygous microsatellite markers in every patient.

Figure 1 shows examples for typical results of LOH analysis with different microsatellite markers. Results of the LOH analyses of all four cystectomies are summarized in table 2 as an overview.

LOH on chromosome 9 was the most frequent event in all neoplastic (papillary tumours and carcinomata in situ) and preneoplastic lesions. In addition, losses on chromosome 9 were found in hyperplasia, dysplasia and carcinomata in situ, as well as in papillary tumours. Losses on chromosome 17p (p53 locus) appeared more often in dysplastic lesions, when compared to other flat lesions of the urothelium (hyperplasia and squamous metaplasia). Interestingly losses on chromosome 8p not only occurred frequently in advanced invasive tumours but also in histologically normal urothelium. Overall, there was a clear correlation between the number of chromosomal losses

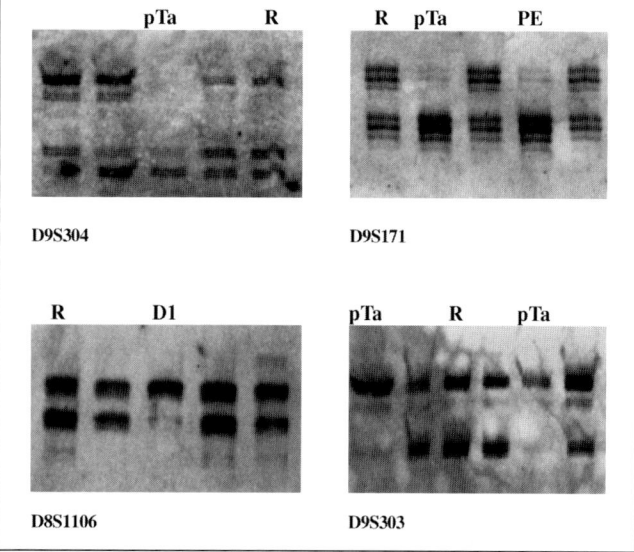

Fig. 1. Representative results of LOH analyses with different microsatellite markers. Loss of one band indicates LOH for this marker. R = DNA from vaginal tissue (control); PE = squamous metaplasia; pTa = superficial papillary tumour; D1 = mild dysplasia, grade 1.

Table 1. Distribution of samples used for the LOH analyses

Histology	Number of samples on		
	chromosome 9	chromosome 8p	chromosome 17p13.1
Squamous metaplasia	9	9	9
Normal urothelium	23	21	22
Hyperplasia	7	5	6
Dysplasia grade 1 (D1)	12	12	12
Dysplasia grade 2 (D2)	6	7	7
Carcinoma in situ	6	6	6
pTa G1–2	5	5	5
pT1G2	2	2	2
pT2G2	2	2	2
Total number	72	69	71

Table 2. Overview of the results of LOH analyses

Histology	Frequency of LOH in different chromosomes				
	9p	9q	9p/9q	8p	17p13.1
Squamous metaplasia	1/9 (11)	–	6/9 (66.6)	2/9 (22.2)	1/9 (11)
Normal urothelium	5/23 (21.7)	4/23 (17.4)	6/23 (26.1)	9/21 (43)	1/22 (4.5)
Hyperplasia	1/7 (14.3)	1/7 (14.3)	4/7 (57.1)	1/5 (20)	1/6 (16.6)
Dysplasia grade 2 (D2)	–	1/6 (16.6)	3/6 (50)	3/7 (42)	2/7 (28.5)
Carcinoma in situ	3/6 (50)	1/6 (16.6)	2/6 (33.3)	3/6 (50)	–
pTaG1–2	–	–	5/5 (100)	–	1/5 (20)
pT1G2	–	–	2/2 (100)	1/2 (50)	1/2 (50)
pT2G2	–	–	2/2 (100)	–	1/2 (50)

Figures in parentheses represent percentage.

and increasing malignancy of the urothelial lesions, especially for losses on chromosome 17p (p53) (fig. 2).

The first cystectomy specimen consisted of a urinary bladder infiltrated by a leiomyosarcoma of the uterus, but without evidence of urothelial lesions. The entire urothelium of this bladder showed a histologically normal urothelium without any sites of dysplasia. LOH analysis revealed no loss in any of the investigated microsatellite markers. No chromosomal alteration was detectable in the urothelium of the non-tumour-bearing bladder.

The cystectomy specimens of patient 2 showed multifocal bladder tumours with variable stages and grades (from pTaG1–2 to pT1G2). Although there was a heterogeneity in the genetic alterations, LOH analysis revealed that these multifocal tumours were monoclonal (fig. 3). The deletion pattern of early losses on chromosome 9, already detectable in histologically normal urothelium, showed a first common loss on 9q21 (D9S303). This first transformed cell cluster could be divided into two different populations with an accumulation of varying chromosomal losses. The first population shared losses on 9q32 (D9S747) and 9p21 (D9S171, D9S304). With additional losses on 8p and 17p13.1 (p53), this population showed progression to papillary and invasive tumours. The second population had common losses on 9p21 (D9S304) but no further detectable chromosomal losses. In addition, this second population showed no malignant progression.

The cystectomies of patients 3 and 4 showed multifocal poorly differentiated, superficial papillary (pTaG3) and muscle invasive (pT2G3) tumours and multifocal carcinomata in situ. LOH analyses revealed a considerable heterogeneity within the genetic alterations of the samples investigated in these patients. Nevertheless, evidence for the oligoclonal development of the neoplastic lesions could be detected. Patient 3 respresents the findings (fig. 4).

The majority of urothelial lesions in patient 3 showed a common loss on 9q34.2 (D9S65). This deletion was detected in lesions ranging from histologically normal urothelium to carcinoma in situ. From this clone two tumour cell populations were separated. One population accumulated losses on 9p21 (pkY3, pkY11). The second population also showed losses on 9p21, however at another region (D9S304). The second clone showed the first LOH on 9p21 (pkY3). This loss was followed by a second LOH on 9p21 (pkY11). Although this second tumour clone also showed chromosomal losses on 9p21, in comparison with the first clone, there was no loss of D9S65 on 9q34.2. This finding of a second, independent clone gave strong evidence for an oligoclonal development of urothelial lesions within the same cystectomy specimen.

Discussion

Methodological Aspects of LOH Analysis with Microsatellite Markers on Microdissected Tissue

LOH analysis is a useful tool for the detection of inactivation of various tumour suppressor genes. According to the hypothesis of Knudson [17], the inactivation of tumour suppressor genes follows a 'two-hit' inactivation. The first hit is the knockout of one allele by a somatic mutation, the second hit is the loss of the remaining allele by a deletion. This deletion of the second allele is accompanied by the LOH.

Up to now most studies investigated only advanced, poorly differentiated tumours in LOH analysis. Conse-

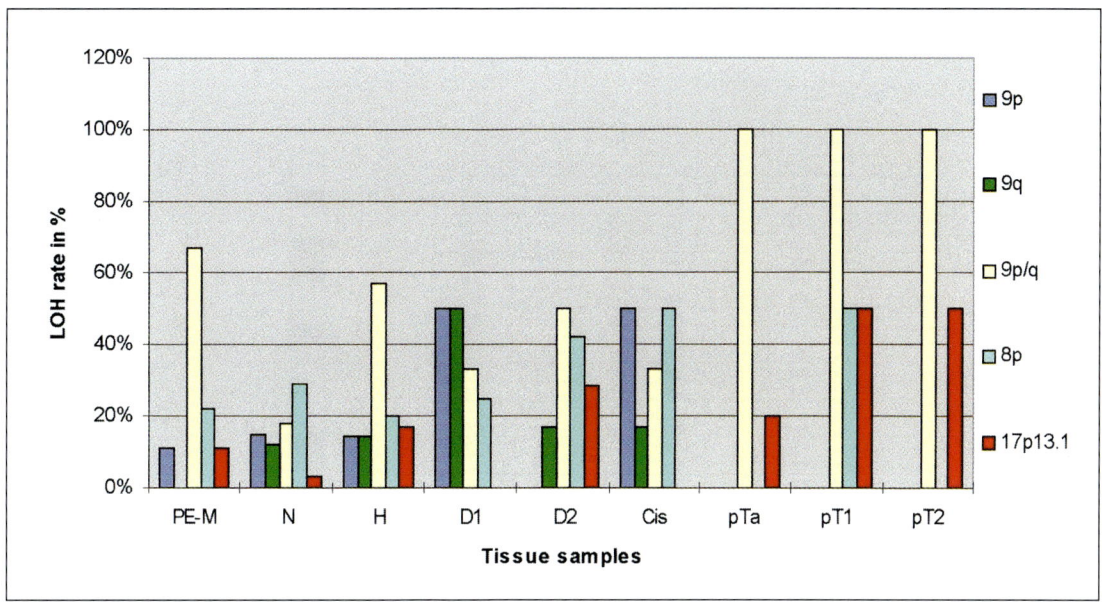

Fig. 2. Summary of the results of all LOH analyses. Data from all investigated samples are shown. PE-M = Squamous metaplasia; N = normal urothelium; H = hyperplasia; D1 = mild dysplasia, grade 1; D2 = moderate dysplasia, grade 2; Cis = carcinoma in situ; pTa = superficial papillary tumour; pT1G2 = invasive tumour; pT2G2 = muscle invasive tumour.

quently very little is known about the exact genetic alterations in preneoplastic lesions and early neoplasias. Particularly with regard to prognosis and early detection of neoplastic transformation, it is important to emphasize the molecular analyses of preneoplasia and histologically normal urothelium.

While microdissection helps to obtain small and distinct morphological entities as the source for DNA extraction for further analysis, careful handling is still required to obtain a pure tumour cell population. Contaminating stromal and inflammatory cells should not exceed 20%, since they cause false-positive LOH findings [18]. Having dissected a small amount of tissue, the next requirement is the representative amplification of DNA. Whole genomic amplification is a valuable method, and I-PEP-PCR proved useful for our work [9]. DNA preamplification can improve the significance of the LOH analysis by enlarging the number of microsatellite markers that can be analyzed. A major problem of the LOH analysis is the appearance of PCR pitfalls like preferential amplifications of alleles. This phenomenon can pretend that there is a loss of an allele or hide informative cases. To avoid these problems every LOH result should be verified in a second independent PCR reaction under the same conditions.

Finally, it has to be discussed whether careful microdissection of small groups of tumour cells does result in pseudoclonality, since: (1) a small group of cells may represent the daughter cells of a proliferating cell still contiguous at the time of analysis, and (2) patch size (defined as the size of tissue originated from a single cell and determined by the number of stem cells per organ) could lead to the false determination of clonality in closely adjacent lesions [19]. For these discussions it is helpful that patch size in urinary bladder has been defined as 120 mm^2 [20]. Problems due to patch size only arise when we are looking at the heterogeneity within a lesion. Sampling different locations with different diagnoses in the bladder mucosa and comparing clonality amongst these lesions avoid the problem of patch size.

Histopathological-Genetic Mapping for New Answers of Urothelial Carcinogenesis

The histologically normal bladder showed no deletions with the markers investigated. Although a number of biopsies have been investigated, it has to be considered that this is material of only one individual, and only documents the evident likelihood of normal urothelium showing negligible amounts of deletions. Baud et al. [21] have used a much higher number of chromosome 9 primers for

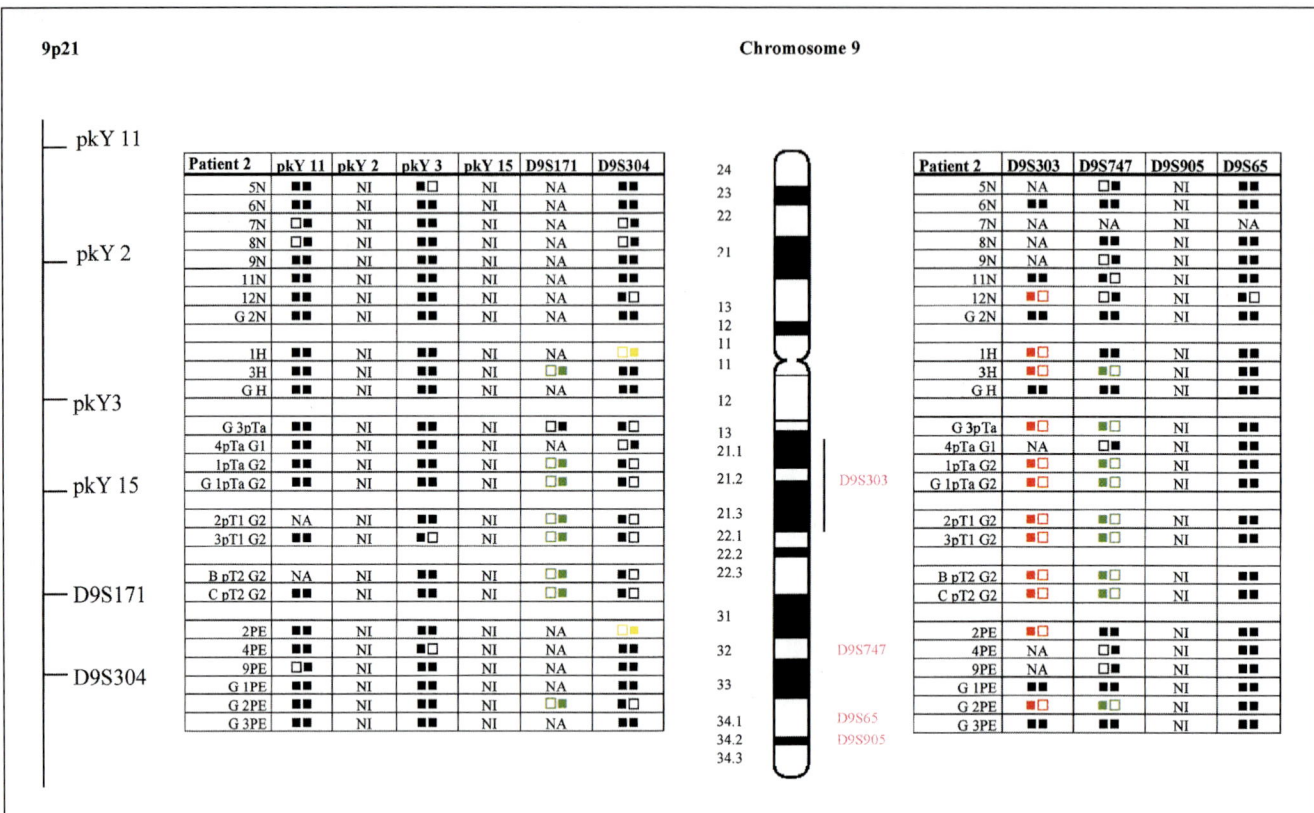

microsatellite analysis, and interestingly found a frequent deletion of one marker in normal urothelium of non-tumour-bearing bladders. Ongoing studies in our laboratory using urothelium of transurethral resections in patients with benign prostate hyperplasia will contribute to the question of changes in normal urothelium.

In contrast to the normal urothelium of the normal bladder, the normal urothelium of tumour-bearing bladders as well as other benign lesions (hyperplasia and squamous metaplasia) already showed deletions with a predominance of chromosomes 8 and 9, and only rare detection of p53 deletion. While seeding of cells from existent tumours into the normal urothelium as the first micrometastatic deposit or intraurothelial tumour cell migration have to be considered as possible explanations for this finding, the detection of initial onset of tumours in histologically benign urothelium or related lesions presents an alternative explanation. Koss [22] already assumed from experience in histomorphological diagnosis that the benign hyperplasia, the simple thickening of the urothelium, is a precursor of papillary neoplasms, and data from our group could confirm this observation by genetic analysis. Hyperplasias analyzed by FISH to detect chromosomes 9 and p53 deletions showed genetic changes in 75% of the cases (n = 14) with findings being identical or less than in papillary tumours of the same patients. Normal urothelium also showed deletions [6]. Other authors have documented aberrant cells in the normal urothelium either by immunohistochemistry [23] or genetic analysis [24].

Fig. 3. LOH pattern of chromosomes 9p and 9q of patient 2. ■ ■ = Heterozygosity; ■ □ = loss of one allele (=LOH); NI = not informative; NA = not available. Red closed and open boxes indicate the first common chromosomal loss on 9q. Green closed and open boxes show the chromosomal losses of the first separated cell population from the initial clone. Yellow open and closed boxes show the chromosomal losses of the second separated population from the initial clone. Monoclonal development of multifocal bladder carcinoma is shown.

Fig. 4. LOH pattern of chromosomes 9p and 9q of patient 3. ■ ■ = Heterozygosity; ■ □ = loss of one allele (LOH); NI = not informative; NA = not available; MSI = microsatellite instability. Red closed and open boxes represent first common chromosomal loss on 9q of the first clone, green open and closed boxes represent chromosomal losses of the first separated population from the first clone, yellow closed and open boxes represent chromosomal losses of the second separated population from the first clone, and blue open and closed boxes represent chromosomal losses on 9p of the second clone. Oligoclonal development of the multifocal bladder carcinoma is shown.

The distribution of frequencies of deletions in table 2 does not clearly separate the entities of flat neoplasia and papillary neoplasia by either chromosome 9 or p53. It is especially unequivocal that significant percentages of chromosome 9 deletions occur in flat neoplasias, raising doubt regarding the simplicity of two clear separate pathways for flat lesions and their consequent solid tumours versus primary papillary tumours [25].

Looking at the individual bladder samples with tumours, patient 2 represents a case without flat urothelial neoplasia and merely papillary tumours. On the basis of a common deletion in normal urothelium and in papillary tumours at the locus 9q21 (D9S303) a clonal expansion with (increasing number of deletions) and without (no other markers affected) clinical progression can be derived and is matched by histology. This is in contrast to the concept assumed by van Tilborg et al. [26] stating that the deletions on chromosome 9 are random events, and a mere indicator of genetic instability. Our finding of clonal divergence with different degrees of accumulation of deletions with tumour progression is also seen for the other two bladders; however, in early lesions two separate deletions are found, indicating two clones. The statement of the existence of two clones is made, since it is well established that chromosome 9 deletions are an early event in bladder cancer [27], thus a common atypical progenitor cell for the two different chromosome 9 deletions is unlikely. However, as long as we deal with deletions as a marker for clonality, we cannot completely exclude a selective exogenous carcinogenic event, causing the deletion. This possibility is best known from aflatoxin causing one type of p53 mutations in hepatocellular carcinoma [28]. Up to now the best tool for studying clonality are methods based on restriction length polymorphisms of X-chromosome-linked genes, as e.g. the androgen receptor gene [29]. The principle of this method is based on the random inactivation of either of the two X-chromosomes early in embryogenesis by methylation of cytosine residues in promoter regions. It is believed that this inactivation is stable even during a neoplastic change. Therefore, we have chosen specimens of female patients for this study to continue work with the comparative analysis of LOH results to X-chromosomal inactivation.

The histological-genetic mapping of bladder cancer provides us with an apt tool to pursue the question of clonality. The finding of oligoclonality has to be substantiated further since it affects the concept of complementary genetic screening for tumour recurrences, which is currently based on the profile of one tumour found in the bladder, assuming monoclonality [30]. The second important insight for the practising urologist gained from mapping in bladder urothelium is derived from genetic changes in benign bladder lesions. The pattern of genetic changes in these lesions can be related to the tumours in the same patients, and questions the possibility of radical tumour resection in standard white light endoscopy.

Acknowledgments

We thank Helmut Kutz, Andrea Schneider and Monika Kerscher for excellent technical support in the areas of tissue processing and LOH analysis. Prof. Dr. f. Hofstädter, Head of the Department, is acknowledged especially for constant support and critical discussion of the research work. The work is supported by grants of the Dr. Mildred Scheel Foundation of Cancer Research (10-1096-Ha1) and the DFG (Kn 263/7-2).

References

1 Sidransky D, Frost P, Von Eschenbach A, Oyasu R, Preisinger AC, Vogelstein B: Clonal origin bladder cancer. N Engl J Med 1992;326: 737–740.
2 Harris AL, Neal DE: Bladder cancer – Field versus clonal origin. N Engl J Med 1992;326: 759–761.
3 Bender CM, Jones PA: Molecular genetics in carcinoma of the bladder; in Petrovich ZBLBL (ed): Carcinoma of the Bladder: Innovations in Management. Heidelberg, Springer, 1999, pp 37–51.
4 Kriegmair M, Baumgartner R, Knuechel R, Stepp H, Hostaedter F, Hofstetter A: Detection of early bladder cancer by 5-aminolevulinic acid induced porphyrin fluorescence. J Urol 1996;155:105–109.
5 Jichlinski P, Forrer M, Mizeret J, Glanzmann T, Braichotte D, Wagnieres G, Zimmer G, Guillou L, Schmidlin F, Graber P, van den Bergh H, Leisinger HJ: Clinical evaluation of a method for detecting superficial surgical transitional cell carcinoma of the bladder by light-induced fluorescence of protoporphyrin IX following the topical application of 5-aminolevulinic acid: Preliminary results. Lasers Surg Med 1997;20:402–408.
6 Hartmann A, Moser K, Kriegmair M, Hofstetter A, Hofstaedter F, Knuechel R: Frequent genetic alterations in simple urothelial hyperplasias of the bladder in patients with papillary urothelial carcinoma. Am J Pathol 1999;154: 721–727.
7 Hartmann A, Rosner U, Schlake G, Dietmaier W, Zaak D, Hofstaedter F, Knuechel R: Clonality and genetic divergence in multifocal low-grade superficial urothelial carcinoma as determined by chromosomes 9 and p53 deletion analysis. Lab Invest 2000;80:709–718.
8 Czerniak B, Chaturvedi V, Li L, Hodges S, Johnston D, Roy JY, Luthra R, Logothetis C, von Eschenbach AC, Grossman HB, Benedict WF, Batsakis JG: Superimposed histologic and genetic mapping of chromosome 9 in progression of human urinary bladder neoplasia. Implications for a genetic model of multistep urothelial carcinogenesis and early detection of urinary bladder cancer. Oncogene 1999;18: 1185–1196.

9 Dietmaier W, Hartmann A, Wallinger S, Heinmoller E, Kerner T, Endl E, Jauch KW, Hofstaedter F, Ruschoff J: Multiple mutation analyses in single tumor cells with improved whole genome amplification. Am J Pathol 1999;154: 83–95.
10 Sobin LH, Wittekind C: TNM Classification of Malignant Tumours. New York, Wiley-Liss, 1997.
11 Murphy WM, Beckwith JB, Farrow GE: Atlas of tumor pathology; in Rosai J, Sobin LH (eds): Tumors of the Kidney, Bladder, and Related Urinary Structures. Washington, Armed Forces Institute of Pathology, 1994.
12 Mostofi FK, Davis CJJ, Sesterhenn IA: Histological typing of urinary bladder tumours; in World Health Organization – International Histological Classification. Berlin, Springer, 1999.
13 Wild P, Knuechel R, Dietmaier W, Hofstaedter F, Hartmann A: Laser microdissection and microsatellite analyses of breast cancer reveal a high degree of tumor heterogeneity. Pathobiology 2001;68:180–190.
14 Schlegel J, Bocker T, Zirngibl H, Hofstaedter F, Ruschoff J: Detection of microsatellite instability in human colorectal carcinomas using a non-radioactive PCR-based screening technique. Virchows Arch 1995;426:223–227.
15 Futreal PA, Barrett JC, Wiseman RW: An Alu polymorphism intragenic to the TP53 gene. Nucleic Acids Res 1991;19:6977.
16 Cairns P, Polascik TJ, Eby Y, Tokino K, Califano J, Merlo A, Mao L, Herath J, Jenkins R, Westra W: Frequency of homozygous deletion at p16/CDKN2 in primary human tumours. Nat Genet 1995;11:210–212.
17 Knudson AG: Antioncogenes and human cancer. Proc Natl Acad Sci USA 1993;90:10914–10921.
18 Bohm M, Kirch H, Otto T, Rubben H, Wieland I: Deletion analysis at the DEL-27, APC and MTS1 loci in bladder cancer: LOH at the DEL-27 locus on 5p13-12 is a prognostic marker of tumor progression. Int J Cancer 1997;74:291–295.
19 Garcia SB, Novelli M, Wright NA: The clonal origin and clonal evolution of epithelial tumours. Int J Exp Pathol 2000;81:89–116.
20 Tsai YC, Simoneau AR, Spruck CH, Nichols PW, Steven K, Buckley JD, Jones PA: Mosaicism in human epithelium: Macroscopic monoclonal patches cover the urothelium. J Urol 1995;153:1697–1700.

21 Baud E, Catilina P, Boiteux JP, Bignon YJ: Human bladder cancers and normal bladder mucosa present the same hot spot of heterozygous chromosome-9 deletion. Int J Cancer 1998;77:821–824.
22 Koss L: Tumors of the urinary bladder; in Atlas of Tumor Pathology. Washington, Armed Forces Institute of Pathology, 1975.
23 Wagner U, Sauter G, Moch H, Novotna H, Epper R, Mihatsch MJ, Waldman FM: Patterns of p53, erbB-2, and EGF-r expression in premalignant lesions of the urinary bladder. Hum Pathol 1995;26:970–978.
24 Muto S, Horie S, Takahashi S, Tomita K, Kitamura T: Genetic and epigenetic alterations in normal bladder epithelium in patients with metachronous bladder cancer. Cancer Res 2000; 60:4021–4025.
25 Spruck CH, Ohneseit PF, Gonzalez-Zulueta M, Esrig D, Miyao N, Tsai YC, Lerner SP, Schmutte C, Yang AS, Cote R: Two molecular pathways to transitional cell carcinoma of the bladder. Cancer Res 1994;54:784–788.
26 van Tilborg AA, Groenfeld LE, van der Kwast TH, Zwarthoff EC: Evidence for two candidate tumour suppressor loci on chromosome 9q in transitional cell carcinoma (TCC) of the bladder but no homozygous deletions in bladder tumour cell lines. Br J Cancer 1999;80:489–494.
27 Simoneau M, Aboulkassim TO, LaRue H, Rousseau F, Fradet Y: Four tumor suppressor loci on chromosome 9q in bladder cancer: Evidence for two novel candidate regions at 9q22.3 and 9q31. Oncogene 1999;18:157–163.
28 Hsu HC, Chiou TJ, Chen JY, Lee CS, Lee PH, Peng SY: Clonality and clonal evolution of hepatocellular carcinoma with multiple nodules. Hepatology 1991;13:923–928.
29 Vogelstein B, Fearon ER, Hamilton SR, Preisinger AC, Willard HF, Michelson AM, Riggs AD, Orkin SH: Clonal analysis using recombinant DNA probes from the X-chromosome. Cancer Res 1987;47:4806–4813.
30 Schneider A, Borgnat S, Lang H, Regine O, Lindner V, Kassem M, Saussine C, Oudet P, Jacqmin D, Gaub MP: Evaluation of microsatellite analysis in urine sediment for diagnosis of bladder cancer. Cancer Res 2000;60:4617–4622.

Detection of Gene Amplification in Intraductal and Infiltrating Breast Cancer by Laser-Assisted Microdissection and Quantitative Real-Time PCR

Sabine Glöckner   Ulrich Lehmann   Nadine Wilke   Wolfram Kleeberger
Florian Länger   Hans Kreipe

Institute of Pathology, Medizinische Hochschule Hannover, Germany

Key Words

Laser microdissection · PCR, real time · Gene amplification · Ductal carcinoma in situ · Breast cancer · CyclinD1 · c-erbB2 · Topoisomerase IIα · c-myc

Abstract

Gene amplification is one essential mechanism leading to oncogene activation which is supposed to play a major role in the pathogenesis of invasive breast cancer. However, using standard methodologies the detection of gene amplifications has been limited especially in small-sized lesions, like pre-invasive precursor lesions. The combination of two novel technologies, laser-based microdissection and quantitative real-time PCR, facilitates the detection of low-level amplifications in morphologically defined lesions. As a model system we investigated in situ breast cancer (ductal carcinoma in situ, DCIS) classified according to the morphology-based Van Nuys grading system for amplification of growth-regulatory genes. In this study 83 formalin-fixed, paraffin-embedded archival DCIS specimens were examined after laser-based microdissection by quantitative real-time PCR using the TaqMan® detection system for amplification of the c-erbB2, topoisomerase IIα, c-myc and cyclinD1 gene. In a subset of 17 DCIS with adjacent infiltrating tumour components we compared intraductal and invasive tumour components in parallel for differences in amplification status. The combination of these new techniques represents an excellent tool to gain new insights into carcinogenesis by analyzing genetic alterations in morphologically identified heterogeneous lesions in breast cancer progression within the very same specimen or even tissue slide.

Copyright © 2001 S. Karger AG, Basel

Introduction

In general, genetic events in carcinogenesis appear to be a combination of activation of oncogenes and loss of suppressor genes. Gene amplification represents an essential mechanism leading to oncogene activation which is a common event in the progression of different human cancers [1, 2].

Breast cancer progression gradually builds up on the accumulation of genetic alterations [3, 4], among which DNA amplifications seem to play a prevalent role concerning about 20 amplified chromosome regions in comparative genomic hybridization (CGH) analyses [5].

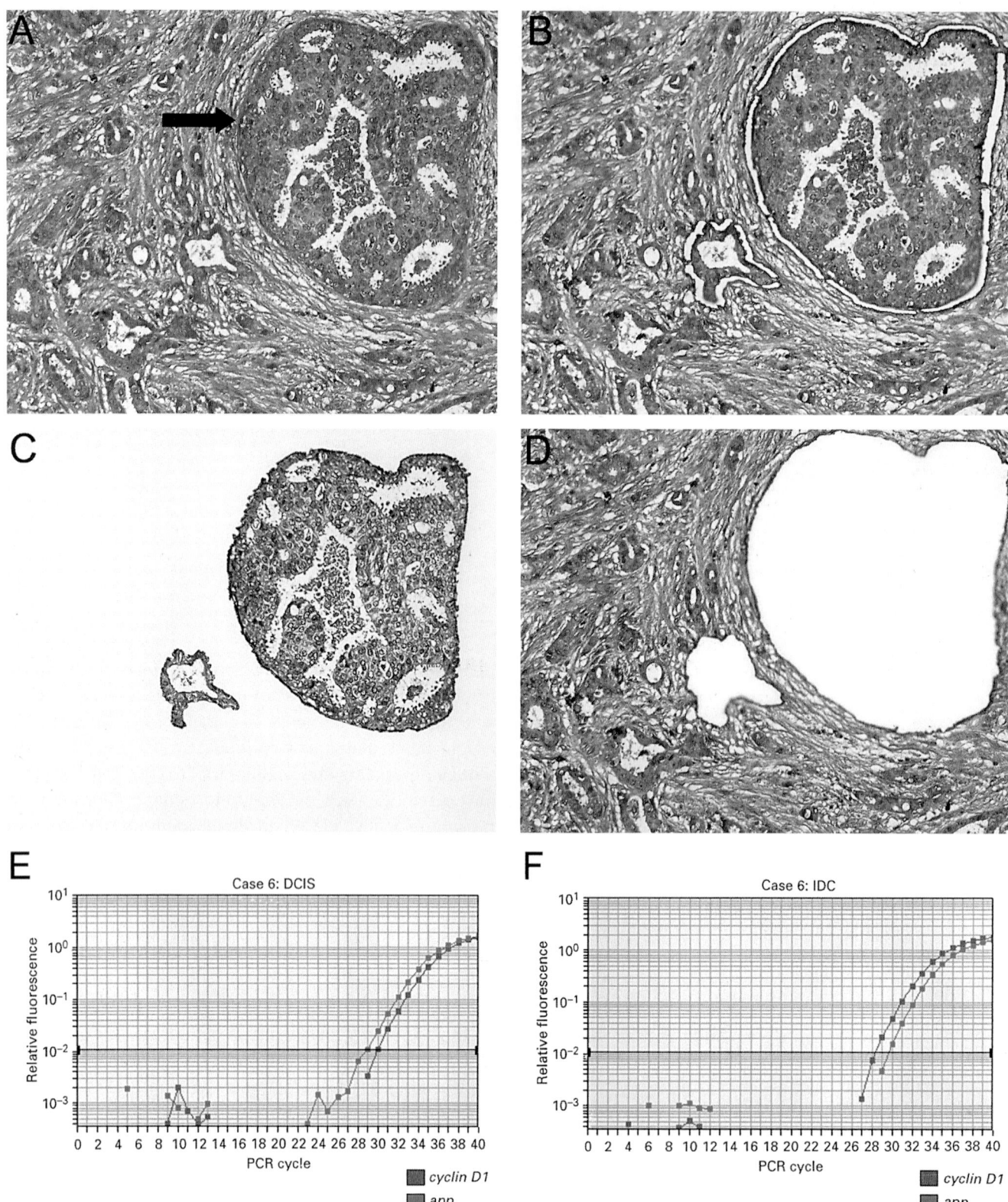

Growth-regulatory genes frequently affected by amplification in breast cancer are the genes for c-*erbB2*, *topoisomerase IIα*, c-*myc* and *cyclinD1* [1]. Some of these amplified genes have been shown to have diagnostic, prognostic and therapeutic relevance [6]. However, the exact and reproducible quantitative detection of gene amplifications has been limited for technical reasons, for example by the admixture of non-neoplastic bystander cells in the necessary crude tumour tissues when using Southern analysis, the often unsatisfying preservation of morphology when using in situ methodologies or due to the relative lack of sensitivity when using CGH analysis.

In this study we combined laser-assisted microdissection and quantitative real-time PCR for detection of gene amplification in paraffin-embedded intraductal and invasive breast cancer specimens [7]. Laser-based microdissection using the PALM® system enables the isolation of morphologically identified pure tumour cell groups for further analysis. This is particularly important in breast carcinoma, which frequently induces marked desmoplastic or inflammatory reaction and is often accompanied by other hyperplastic or neoplastic proliferative lesions. The subsequent quantitative PCR analysis of the dissected cell groups with the TaqMan® detection system allows the objective and reproducible quantification of low-level amplifications in minute tissue samples [7–9]. The combination of these two novel technologies provides an excellent tool for studies of genetic alterations in neoplastic breast lesions as well as for comparison of different or heterogeneous neoplastic lesions, i.e. for example intraductal and adjacent invasive components, in one specimen or even tissue slide (fig. 1A–D).

Breast cancer as well as ductal carcinoma in situ (DCIS) as its potential precursor lesion represent a highly heterogeneous disease, and evidence is mounting that this heterogeneity finds its source in genetic variability [3, 10]. With respect to phenotypic diversity, various morphological classification schemes for DCIS have been proposed in recent years [11–13]. Among these classification proposals, the Van Nuys (VN) grading system facilitates categorization of DCIS by grouping it into three grades based on nuclear size and pleomorphism and the presence or absence of necrosis. However, in common with other histopathological grading protocols interobserver reproducibility of this simplified system is also limited due to difficulties concerning standardization and independence of subjective influences [14, 15], which calls for tumour classifications based on the identification of novel biological and molecular markers. Reflecting the need for a more objective basis of DCIS classification, idealistic 'future' classifications aim at combining phenotypic aspects with hallmark genetic alterations in order to improve diagnostic reproducibility [16].

For correlation of genotype with phenotype, we tested DCIS lesions classified according to the morphology-based VN grading system for the amplification of growth-regulatory genes (c-*myc*, c-*erbB2*, *topoisomerase IIα*, *cyclinD1* gene) that are implicated in breast cancer development. Moreover, for comparison of amplification status at different steps of breast cancer progression we tested intraductal and adjacent infiltrating tumour cell clones in parallel in a subset of 17 tumours.

Materials and Methods

Ductal Carcinoma in situ

83 formalin-fixed, paraffin-embedded DCIS specimens were retrieved from the archive of the Institute of Pathology of the Hannover Medical School. One representative formalin-fixed, paraffin-embedded block was chosen for analyses from each case after review of haematoxylin and eosin-stained slides. Among the total of 83 cases included in this study, 37 cases were diagnosed as pure DCIS without invasive carcinoma development according to a TNM stage of pTis. The remaining 46 cases revealed intraductal and adjacent invasive tumour components corresponding to a TNM stage of pT1–2. Among these, 17 cases were chosen for comparison of amplification status in intraductal and infiltrating tumour components.

Classification of DCIS

Each DCIS case was independently examined by two pathologists (H.K. and F.L.) and categorized into high (VN3), intermediate (VN2) and low-grade (VN1) according to the three-tiered system of the VN classification [13]. Discrepant cases were identified and reviewed together on a multiheaded microscope in order to achieve consensus. Group VN1 consisted of 11 DCIS cases (13%) classified as low grade, group VN2 comprised 12 intermediate-grade DCIS (14%), and 60 DCIS cases (72%) were included in high-grade group VN3 (table 1).

Fig. 1. Laser-assisted microdissection and TaqMan PCR of isolated intraductal and adjacent invasive breast cancer cell groups. **A–D** Isolation of intraductal (arrow) and adjacent infiltrating tumour cell populations of a ductal invasive carcinoma using the laser pressure catapulting technique. Methylene blue staining. Original magnification × 200. **E, F** Amplification plots of microdissected intraductal (**E**) and infiltrating (**F**) tumour components of a ductal invasive carcinoma, stage pT2 (case 6). *CyclinD1* amplification is detected in ductal invasive tumour components (IDC in **F**) by shifting of the amplification plot of the *cyclinD1* gene to the left (low-level amplification: 2.1 ± 0.1), in comparison to isolated intraductal tumour components (DCIS in **E**) of the same specimen showing no *cyclinD1* amplification. *app* gene serves as internal reference gene in both PCR analyses.

Table 1. Frequency of gene amplifications in 83 DCIS cases in correlation with grading and the absence (pTis) or presence (pT1–pT2) of invasive tumour components

Amplification	All DCIS	VN1	VN2	VN3	pTis	pT1–2
c-erbB2	21	–	1	20	12	9
Topoisomerase IIα	3	–	–	3	2	1
c-myc	5	–	1	4	1	4
CyclinD1	5	–	–	5	4	1
n	83	11	12	60	37	46

Microdissection

3- to 5-μm tissue sections were cut from the selected paraffin block and mounted on polylysine-coated polyethylene foil. Sections were dewaxed and rehydrated following standard protocols. After methylene blue staining microdissection of the tissue sections was performed by using the PALM Laser-MicroBeam System (PALM, Wolfratshausen, Germany), which enables the contact-free isolation of selected lesions by means of the laser pressure catapulting technique into the lid of a 0.5-ml reaction tube (fig. 1A–D). DNA was isolated by the addition of 40 μl of proteinase K digestion buffer into the lid and overnight incubation of this mixture in a hybridization oven at 40°C. After heat inactivation of proteinase K an aliquot of this lysate was used for subsequent PCR analysis.

PCR, Probes and Primers

TaqMan PCR was performed essentially as previously described [7] using a 30-μl final reaction mixture containing 250 nmol/l of each primer, 150 nmol/l probe, 1 unit of AmpliTaq Gold, 200 μmol/l each of dATP, dCTP, dTTP, dGTP in 10× TaqMan buffer A. The reaction mixture was preheated at 95°C for 10 min, followed by 40 cycles at 95°C for 15 s and 60°C for 1 min. The primer and probe sequences and the magnesium concentration for each gene can be obtained from the corresponding author.

Evaluation of Results

The relative gene copy number was evaluated on the basis of the threshold cycles (C_T values) of the gene of interest and of an internal reference gene. As internal reference gene *app* was selected, sited on locus 4q11–q13, for which no amplifications in breast cancer have been reported. By subtraction of the C_T value of the internal reference gene from the C_T values of test genes in the same tumour sample the relative C_T values, i.e. $\Delta C_{T(DCIS)}$ values, for each of the four target genes were determined. These $\Delta C_{T(DCIS)}$ values for each test gene were then compared with a reference range derived from $\Delta C_{T(normal\ tissue)}$ values. To determine this reference range, the PCR results of sixty microdissected non-neoplastic tissue samples (mostly stroma cells, lymphocytes or vessels) were filed, derived from slides of DCIS cases included in this study with adjacent 'normal' tissue on the selected paraffin block. For elimination of PCR variabilities due to fixation artefacts or sample impurities the reference range for each test/reference gene pair in normal tissue was determined as follows:

reference range = [mean ($\Delta C_{T(normal\ tissue)}$) ± σ_{n-1}] ± 1.

The reduction of the relative C_T or ΔC_T value by one cycle is equivalent to a level of amplification of 1.8 ± 0.1. Consequently, amplification of a test gene leads to a smaller $\Delta C_{T(DCIS)}$ value below the reference range limit. The level of amplification was graded as follows: low-level amplification with copy number increase ranging from 2- to ≤5-fold, intermediate-level amplification from >5- to ≤10-fold and high-level amplification with >10-fold.

Statistical Analysis

For the purpose of statistical evaluation, results were tabulated and analyzed with the linear trend test (exact) and Kruskal-Wallis test. p values < 0.05 were considered statistically significant.

Results

Gene Amplifications in DCIS

The results of the PCR analyses of all DCIS specimens as well as distribution of amplifications in different grades according to VN classification are summarized in table 1. In total, 30 out of 83 (36%) DCIS specimens revealed gene amplifications: c-*erbB2* amplification represented the most frequent copy number increase of all genes under study detected in 21 out of 83 (25%) cases. 6 of these specimens exhibited low-level amplification (2- to 5-fold), 11 cases presented intermediate-level amplification (>5- to 10-fold). The remaining 4 cases presented high-level c-*erbB2* amplifications with a copy number increase up to 25-fold (fig. 2).

Topoisomerase IIα amplification was detected in 3 out of 83 (4%) cases. All these specimens revealed low-level amplification. 5 out of 83 (6%) cases exhibited low-level c-*myc* amplification with amplification factors between 3- and 4-fold.

CyclinD1 amplification was found in 5 (6%) specimens. 4 cases presented low-level amplification and only 1 case presented intermediate-level amplification with 6-fold copy number increase.

Gene Amplifications and VN Classification

The distribution of gene amplification in the different VN groups is shown in table 1. In group VN1 comprising 11 cases no copy number aberrations of the genes under study were detected. 2 specimens among the 12 cases in group VN2 revealed gene amplifications: 1 case showed low-level c-*erbB2* amplification (4-fold), another case low-level c-*myc* amplification (4-fold). In group VN3, including 60 cases, 28 specimens showed amplifications of the genes under study. The most frequent amplification concerned c-*erbB2* detected in 21 cases (25%). Statistical analysis revealed a highly significant correlation between high-grade DCIS VN3 and frequency of amplification in the exact linear trend test (p = 0.006) as well as the amplification level in Kruskal-Wallis test (p = 0.02). *Topoisomerase IIα* amplifications occurred in three high-grade DCIS: two presented a combined c-*erbB2*/*topoisomerase IIα* co-amplification, the remaining case a solitary *topoisomerase IIα* copy number increase. c-*myc* amplification was present in 4 high-grade DCIS specimens. *CyclinD1* amplification occurred exclusively in group VN3 in a subset of 5 cases. 1 case in this group presented a composite amplification of c-*erbB2*, c-*myc* as well as *cyclinD1* gene, representing three target genes at different loci. The distribution of gene amplifications exhibited a highly significant correlation to grading according to VN classification in the exact linear trend test (p = 0.003).

Gene Amplifications in Intraductal and Adjacent Invasive Tumour Components

In a subset of 17 specimens, intraductal and adjacent infiltrating tumour components were microdissected and analyzed in parallel for comparison of amplification status (fig. 1). The results of the PCR analyses are listed in table 2. Among these specimens, 16 presented the same amplification status in terms of presence or absence of gene amplifications: in 15 cases neither the intraductal nor the infiltrating tumour components presented amplifications of the genes under study, while another case presented c-*myc* amplification in both components. In one tumour only (case 6) a divergent amplification status was found with presentation of *cyclinD1* amplification exclusively in the invasive ductal carcinoma components (fig. 1E, F).

Discussion

By means of laser-based microdissection it is now possible to gain new insights into cancer progression by analyzing genetic alterations in morphologically defined cells at different stages of disease progression. By preserving cytological and tissue architecture necessary to confirm histopathological criteria in early epithelial lesions this

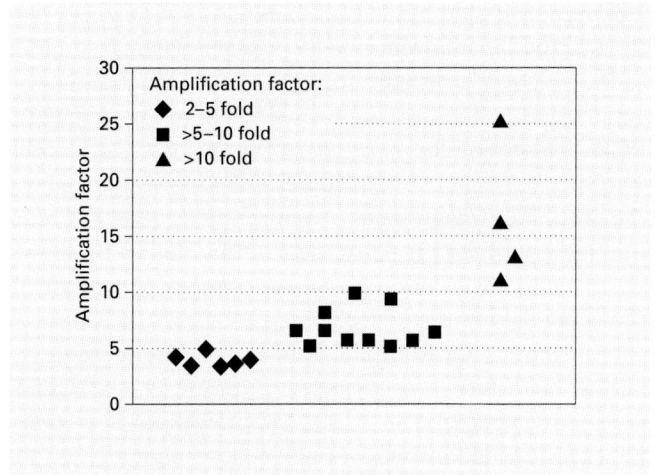

Fig. 2. Amplification level of c-*erbB2* gene in 21 DCIS specimens. Each of the symbols corresponds to 1 case. 6 cases present low-level amplification with amplification factors between 2- and 5-fold, 11 cases intermediate-level amplification with amplification levels between >5- and 10-fold and the remaining 4 cases high-level amplifications with amplification factors above 10-fold.

Table 2. Comparison of amplification status in intraductal and infiltrating tumour components of the same specimen in a subset of 17 DCIS cases of stage pT1–2

No.	Amplification status (intraductal/infiltrating)
06	–/*cyclinD1*↑
08	–/–
09	–/–
17	–/–
26	c-*myc*↑/c-*myc*↑
31	–/–
58	–/–
59	–/–
72	–/–
78	–/–
79	–/–
80	–/–
81	–/–
83	–/–
84	–/–
91	–/–
92	–/–

↑ = Amplified.

technique allows the isolation of pure tumour cell populations without contamination by non-neoplastic bystander cells, which is especially important for investigation of small-sized lesions of interest. The subsequent kinetic PCR analysis of the isolated cell populations using the TaqMan detection system represents a powerful tool for the accurate measurement of gene copy numbers in minute tissue samples [7]. Thus the combination of these two novel technologies provides a new approach to comparing the genotypes in distinct phenotypes along the course of carcinogenesis for example from intraductal to invasive and to metastatic breast cancer. In order to relate genotype and phenotype in breast carcinogenesis we tested DCIS lesions for amplification of genes frequently amplified in invasive breast cancer, i.e. c-*erbB2*, *topoisomerase IIα*, c-*myc* and *cyclinD1* [9]. Investigating DCIS as potential precursor lesions of invasive carcinoma might elucidate at which stage of tumour progression these common amplifications occur. Moreover, we applied these techniques for evaluation of an underlying molecular basis – in terms of gene amplifications – of the morphology-based VN classification [17]. As we found amplifications of all tested genes in DCIS as potential precursor lesion of invasive carcinoma, copy number increases of the genes under study might represent an early step in carcinogenesis. Especially c-*erbB2* amplification as the most frequently amplified gene in our study may play an important role in tumour initiation.

The comparably low frequencies of c-*myc* as well as *cyclinD1* amplifications in only 6% (5 of 83 each) of the tested DCIS specimens in comparison to studies performed on invasive carcinomas [18–20] suggest that amplifications of these genes might not represent a decisive genetic alteration in early steps of carcinogenesis. Considering the almost exclusive occurrence of these gene amplifications in high-grade DCIS lesions, overexpression of their gene product might promote tumour progression in later stages of carcinogenesis. Moreover, in keeping with previous studies using different methodologies [21, 22] high-grade DCIS (VN3) show accumulation of genetic alterations in terms of co-amplification and composite amplification, suggesting a far advanced malignant lesion. Some of these alterations might indicate merely an increase of genetic instability during clonal evolution of tumour cells.

Statistical analyses show significant association between distribution of gene amplifications and VN grading underpinning an underlying molecular basis of the morphology-based VN grading system in terms of gene amplifications. Among the four target genes, the frequency and the level of c-*erbB2* amplification differs significantly in group VN3 compared to low- or intermediate-grade DCIS of group VN1 and VN2.

Comparing amplification status in pure in situ intraductal carcinoma (pTis) with DCIS with adjacent infiltrating components (pT1–2) no significant differences were detected. This might imply that amplifications of the genes under study may not be the decisive genetic alteration determining further invasive potential of an intraductal cell clone. When comparing intraductal and infiltrating tumour components in the same specimen only one out of 17 specimens revealed a difference in view of the amplification status, namely the acquisition of *cyclinD1* amplification in the infiltrating tumour component. The very frequent similarity between intraductal and infiltrating tumour components in terms of absence or presence of gene copy number increases may point at a close genetic relationship between the infiltrating and invasive cell clones. Therefore, the results of this study might support the hypothesis that the biological potential of invasive breast cancer may be determined at a preinvasive stage [23, 24]. However, it is not certain that the intraductal components adjacent to invasive tumour components represent the residues of an intraductal progenitor lesion. In fact, intraductal tumour components could also represent an intraductal offspring of an invasive cell clone growing along ducts.

Future studies examining larger case numbers have to show if *cyclinD1* and c-*myc* amplification have an impact on further invasive potential of an intraductal cell clone. These further studies using new techniques like laser microdissection and real-time PCR may find some other genetic differences between intraductal and infiltrating components of breast neoplasia influencing invasive and metastatic potential.

References

1 Knuutila S, Bjorkqvist AM, Autio K, Tarkkanen M, Wolf M, Monni O, Szymanska J, Larramendy ML, Tapper J, Pere H, El-Rifai W, Hemmer S, Wasenius VM, Vidgren V, Zhu Y: DNA copy number amplifications in human neoplasms: Review of comparative genomic hybridization studies. Am J Pathol 1998;152: 1107–1123.

2 Schwab M: Amplification of oncogenes in human cancer cells. Bioessays 1998;20:473–479.

3 Beckmann MW, Niederacher D, Schnurch HG, Gusterson BA, Bender HG: Multistep carcinogenesis of breast cancer and tumour heterogeneity. J Mol Med 1997;75:429–439.

4 Bergstein I: Molecular alterations in breast cancer; in Bowcock AM (ed): Breast Cancer: Molecular Genetics, Pathogenesis, and Therapeutics. Totowa, Humana Press, 1999, pp 143–170.

5 Courjal F, Cuny M, Simony-Lafontaine J, Louason G, Speiser P, Zeillinger R, Rodriguez C, Theillet C: Mapping of DNA amplifications at 15 chromosomal localizations in 1875 breast tumors: Definition of phenotypic groups. Cancer Res 1997;57:4360–4367.

6 Jarvinen TA, Tanner M, Rantanen V, Barlund M, Borg A, Grenman S, Isola J: Amplification and deletion of topoisomerase IIalpha associate with ErbB-2 amplification and affect sensitivity to topoisomerase II inhibitor doxorubicin in breast cancer. Am J Pathol 2000;156:839–847.

7 Lehmann U, Glöckner S, Kleeberger W, Feist H, von Wasielewski R, Kreipe H: Detection of gene amplification in archival breast cancer specimens by laser-assisted microdissection and quantitative real-time polymerase chain reaction. Am J Pathol 2000;156:1855–1864.

8 Heid CA, Stevens J, Livak KJ, Williams PM: Real time quantitative PCR. Genome Res 1996;6:986–994.

9 Bieche I, Olivi M, Champeme MH, Vidaud D, Lidereau R, Vidaud M: Novel approach to quantitative polymerase chain reaction using real-time detection: Application to the detection of gene amplification in breast cancer. Int J Cancer 1998;78:661–666.

10 Cuny M, Kramar A, Courjal F, Johannsdottir V, Iacopetta B, Fontaine H, Grenier J, Culine S, Theillet C: Relating genotype and phenotype in breast cancer: An analysis of the prognostic significance of amplification at eight different genes or loci and of p53 mutations. Cancer Res 2000;60:1077–1083.

11 Holland R, Peterse JL, Millis RR, Eusebi V, Faverly D, van de Vijver MJ, Zafrani B: Ductal carcinoma in situ: A proposal for a new classification. Semin Diagn Pathol 1994;11:167–180.

12 Lagios MD: Grading of ductal carcinoma in situ; in Silverstein MJ (ed): Ductal Carcinoma in situ of the Breast. Baltimore, Williams & Wilkins, 1997, pp 227–231.

13 Silverstein MJ, Poller DN, Waisman JR, Colburn WJ, Barth A, Gierson ED, Lewinsky B, Gamagami P, Slamon DJ: Prognostic classification of breast ductal carcinoma-in-situ. Lancet 1995;345:1154–1157.

14 Bethwaite P, Smith N, Delahunt B, Kenwright D: Reproducibility of new classification schemes for the pathology of ductal carcinoma in situ of the breast. J Clin Pathol 1998;51: 450–454.

15 Wells WA, Carney PA, Eliassen MS, Grove MR, Tosteson AN: Pathologists' agreement with experts and reproducibility of breast ductal carcinoma-in-situ classification schemes. Am J Surg Pathol 2000;24:651–659.

16 Ellis IO, Pinder SE, Lee AH, Elston CW: A critical appraisal of existing classification systems of epithelial hyperplasia and in situ neoplasia of the breast with proposals for future methods of categorization: Where are we going? Semin Diagn Pathol 1999;16:202–208.

17 Silverstein MJ: Van Nuys ductal carcinoma in situ classification; in Silverstein MJ (ed): Ductal Carcinoma in situ of the Breast. Baltimore, Williams & Wilkins, 1997, pp 247–256.

18 Kreipe H, Feist H, Fischer L, Felgner J, Heidorn K, Mettler L, Parwaresch R: Amplification of c-myc but not of c-erbB-2 is associated with high proliferative capacity in breast cancer. Cancer Res 1993;53:1956–1961.

19 Courjal F, Louason G, Speiser P, Katsaros D, Zeillinger R, Theillet C: Cyclin gene amplification and overexpression in breast and ovarian cancers: Evidence for the selection of cyclin D1 in breast and cyclin E in ovarian tumors. Int J Cancer 1996;69:247–253.

20 Dickson C, Fantl V, Gillett C, Brookes S, Bartek J, Smith R, Fisher C, Barnes D, Peters G: Amplification of chromosome band 11q13 and a role for cyclin D1 in human breast cancer. Cancer Lett 1995;90:43–50.

21 Moore E, Magee H, Coyne J, Gorey T, Dervan PA: Widespread chromosomal abnormalities in high-grade ductal carcinoma in situ of the breast. Comparative genomic hybridization study of pure high-grade DCIS. J Pathol 1999; 187:403–409.

22 James LA, Mitchell EL, Menasce L, Varley JM: Comparative genomic hybridisation of ductal carcinoma in situ of the breast: Identification of regions of DNA amplification and deletion in common with invasive breast carcinoma. Oncogene 1997;14:1059–1065.

23 Buerger H, Otterbach F, Simon R, Schafer KL, Poremba C, Diallo R, Brinkschmidt C, Dockhorn-Dworniczak B, Boecker W: Different genetic pathways in the evolution of invasive breast cancer are associated with distinct morphological subtypes. J Pathol 1999;189:521–526.

24 Gupta SK, Douglas-Jones AG, Fenn N, Morgan JM, Mansel RE: The clinical behavior of breast carcinoma is probably determined at the preinvasive stage (ductal carcinoma in situ). Cancer 1997;80:1740–1745.

Laser Microdissection and Microsatellite Analyses of Breast Cancer Reveal a High Degree of Tumor Heterogeneity

Peter Wild   Ruth Knuechel   Wolfgang Dietmaier   Ferdinand Hofstaedter
Arndt Hartmann

Institute of Pathology, University of Regensburg, Germany

Key Words
Breast cancer · Microsatellite instability · Loss of heterozygosity analysis · Laser microdissection · Tumor heterogeneity

Abstract
Carcinomas with productive fibrosis are the most common forms of breast cancer. Analysis of tumor-specific genomic alterations can be compromised by the presence of normal cells, demanding microdissection of small tumor areas to detect loss of heterozygosity (LOH) and microsatellite instability (MSI). The aim of this study was to evaluate the importance of precise laser microdissection for microsatellite analyses and investigation of tumor heterogeneity in breast cancer. 39 primary breast tumor samples were analyzed for MSI and LOH by PCR followed by polyacrylamide gel electrophoresis and silver staining using 15 microsatellite markers. Different tumor areas were processed separately in 30 patients. Both intraductal and invasive breast cancer regions were investigated in 11 patients. The following results were obtained: (1) accurate microdissection revealed MSI in 3 or more of the investigated markers (≥20%) in 33% of the patients, a higher frequency than reported previously; (2) laser microdissection was 43% more sensitive in detection of LOH compared to manual microdissection due to a reduction of contamination by normal cells, and (3) 29 of 30 investigated tumors showed heterogeneity of genetic alterations in different tumor regions. Laser-based microdissection is a valuable tool in genetic analysis of desmoplastic tumors and allows an accurate determination of genetic alterations in histologically different tumor regions.

Copyright © 2001 S. Karger AG, Basel

Introduction

Molecular analysis in tumor pathology should be performed in precisely determined areas with homogenous tumor cells. Accurate selection and molecular investigation of specific tumor areas or preneoplastic lesions could reveal insight into the multistep development of cancer and the heterogeneity of manifest tumors. The analysis of tumor-specific genetic alterations can be compromised by the presence of normal cells. Thus, contamination by stromal and inflammatory cells should be minimal. In order to obtain high-quality tumor DNA from often small tissue samples, accurate tissue microdissection is one of the most useful techniques, becoming increasingly important in molecular pathology. Under microscopic control, histologically heterogenous tumor areas can be exactly defined and dissected from the surrounding nonneoplastic tissue.

The spectrum of techniques ranges from manual microdissection or the use of a micromanipulator to laser-assisted microdissection systems. Laser microdissection is a novel technique in molecular oncology [1-9] and allows the contamination-free isolation of morphologically defined pure cell populations from archival tissue sections.

The use of these new microdissection methods is especially important in carcinomas with productive fibrosis as commonly found in breast cancer. In most breast cancers, there are considerable amounts of stromal and inflammatory cells. For reliable microsatellite analyses and detection of chromosomal deletions by loss of heterozygosity (LOH) studies, a tumor cell content of at least 70-80% is required [10]. The aim of the present study was the evaluation of different microdissection techniques as a prerequisite for the detection of microsatellite instability (MSI) and LOH in breast cancer. Furthermore, the genetic heterogeneity of histopathologically diverse areas of breast cancer was investigated.

In a retrospective study, normal reference tissue and primary breast cancer samples form histologically diverse tumor areas of 39 patients with metastatic breast cancer were microdissected either by laser-assisted microdissection or manual microdissection and analyzed for MSI and LOH by PCR followed by polyacrylamide gel electrophoresis and silver staining using 15 microsatellite markers. The results of the study showed that laser microdissection increased the sensitivity of detection of LOH in breast cancer. Furthermore, accurate microdissection of different tumor regions revealed a heterogeneity of the genetic alterations in almost all investigated tumors.

Materials and Methods

Tumors

All patients were selected from an ongoing prospective study of high-dose chemotherapy followed by autologous stem cell transplantation at the Department of Internal Medicine I, University of Regensburg, Germany. Clinical and histopathological data of all patients are summarized in table 1. Patients had either advanced local breast cancer with excessive lymph node metastases or systemic metastases. Tissue samples of the primary tumor and histopathological diagnoses were obtained from eleven different pathology departments. In 14 cases, only archival hematoxylin-eosin (HE)-stained tissue sections were available for microdissection. In the remaining cases, formalin-fixed and paraffin-embedded tissue was used. In each case normal cells either from normal breast tissue distant from the tumor, or from skin, bone marrow or gastric mucosa were available as reference for microsatellite analyses. Tumor cells of histopathologically heterogenous areas in each tumor, i.e. invasive and intraductal components, were processed separately in 11 patients. In addition, different areas of invasive carcinoma were microdissected and investigated separately in 19 patients. Lymph node and systemic metastases were analyzed if tissue was available (n = 13).

Microdissection

Genomic DNA was prepared from 5-μm-thick paraffin sections after deparaffinization and microdissection. In brief, histological sections were deparaffinized by a 1-hour step at 65°C and incubation in xylene at room temperature for 2 × 15 min. The sections were rehydrated in ethanol (100, 96, 70%; at least 2 × 5 min each). After incubation for 5 min in H_2O, specimens were stained with 0.1% methylene blue and used for microdissection.

Pure tumor cell populations were obtained using either manual microdissection with a needle under an inverted microscope (× 40) or laser microdissection (PALM® Robot Microbeam, PALM, Wolfratshausen, Germany). Methylene blue-stained paraffin-embedded tissue sections as well as archival HE-stained slides were used for microdissection (fig. 1). The results of both microdissection techniques were compared. The aim of the manual microdissection was to obtain at least 70% tumor cells for the molecular analyses. For laser microdissection, the specimens were mounted onto a 1.35-μm polyethylene membrane, which was taped onto a supporting object slide (Tesa® house & garden universal) and processed as previously described [4]. The selected tumor region was cleared from nontumorous cells (e.g. inflammatory cells in areas of intratumoral necrosis) and microdissected precisely following its irregular shape. With one single laser shot the entire microdissected membrane tissue area was ejected and catapulted towards the collector. The adjacent tissue remained entirely intact for further examinations. The catapulted specimen at the collector was morphologically well preserved and was captured in the lid of the reaction tube for documentation of the cell input in the genetic analysis (fig. 1).

In 4 cases, additionally frozen specimens were available for the preparation of frozen sections and touch preparations. 5-μm frozen sections were microdissected after methylene blue staining. Touch preparations from the partially thawed cut surface of frozen tissue were made as described previously [12]. Touch preparations can reveal, like laser microdissection, a pure tumor cell population with minimal contaminating nontumorous cells [12].

HE-stained archival slides were incubated in xylene overnight. After removal of the coverslip, the slides were rehydrated in 96% ethanol but not exposed to 70% ethanol to preserve HE staining. Laser-mediated removal of contaminating normal cells was followed by manual picking of tumor cells with a micromanipulator.

From each specimen at least 100 cells were prepared. Samples were transferred into 1.5-ml tubes containing 180 μl of ATL buffer provided with the DNA isolation kit.

DNA Isolation

For genetic analysis, normal and tumor DNA were extracted using the QIAamp DNA Mini Kit (Qiagen, Hilden, Germany) according to the provided protocol. To increase DNA yield, elution with 2 × 100 μl of 70°C preheated water was performed. Each elution step included a 5-min incubation of the QIAamp spin column with the preheated water at 70°C before centrifugation. After purification PCR template concentration was increased by (1) reducing the elution volume to 50 μl (SpeedVac SC110, Savant, Farmingdale, N.Y.) and (2) whole genome amplification [13] using an improved primer extension preamplification PCR as described previously [14].

Table 1. Clinical and histopathological data, results of microsatellite analyses and investigation of tumor heterogeneity of 39 investigated breast cancer patients

No.	Age[1] years	T	N	M	G	Histology	Locali-zation	Lymph nodes pos./total	ER[2]	PR[3]	Meno-pausal status	Material	MSI per tumor	MSI status[4]	LOH per tumor	CIN status[5]	Tumor hetero-geneity[6]
1	55	2(2)	2	1	3	ductal	ri	14/21	0	0	post	PET,FT,TP	0	0	2	0	1
2	50	4d	1 biii	0	3	ductal	le	5/18	1	1	post	PET,FT,TP	2	1	5	0	1
3	41	1a(m)	2	1	3	ductal	le	23/23	0	0	pre	HE	2	1	4	0	1
4	43	4d	1	0	2	ductal	le	4/17	0	0	pre[7]	PET	2	1	4	0	1
5	48	4d	1	0	3	ductal	le	11/30	0	0	pre	PET	0	0	6	1	1
6	25	2	1	x	3	medullary	le	5/15	0	0	pre	PET	2	1	7	1	1
7	39	1c	x	1	2–3	lobuloductal	le	x	1	1	pre	PET	4	2	6	1	1
8	33	1	1	1	2–3	ductal	le	x	0	0	pre	HE	2	1	7	1	2
9	39	2	1	0	2	ductal	ri	14/15	(1)	(1)	pre	PET	9	2	4	0	1
10	38	1–2	0	0	2–3	ductal	le	0/15	1	0	pre	HE	2	1	5	0	1
11	32	2	2	0	3	ductal	ri	21/26	1	1	pre	PET	6	2	7	1	1
12	32	2	1	0	2	ductal-tubular	le	1/18	1	1	pre	PET	2	1	7	1	1
13	39	2(m)	1	0	2	ductal	ri	1/10	0	0	pre	HE	0	0	2	0	1
14	40	1c	x	1	2	ductal	ri	x	0	1	pre	PET,FT,TP	3	2	1	0	1
15	40	3	1	x	2–3	ductal	le	1/13	0	0	pre	PET	1	1	4	0	1
16	36	2a	1	0	3	ductal	le	5/27	1	1	pre	HE	1	1	1	0	2
17	45	2	1 bii	1	2	lobular	le	11/19	1	1	pre	PET	1	1	4	0	1
18	45	1c	0	1	2	ductal	ri	0/10	0	0	pre	HE	0	0	4	0	2
19	44	2	1	x	3	ductal	le	3/11	1	1	pre	HE	1	1	2	0	1
20	53	2(m)	0	1	2	ductal	le	0/17	1	1	post	PET	1	1	6	1	2
21	47	2	0	0	3	ductal	ri	0/6	0	1	post	HE	2	1	7	1	1
22	31	1b	0	0	3	ductal	le	0/8	0	0	pre	HE	7	2	4	0	1
23	43	2	2	x	3	ductal	le	23/26	0	0	pre	PET	3	2	3	0	2
24	53	4d	2	0	3	ductal	ri	14/16	1	1	post	PET,FT,TP	1	1	1	0	1
25	41	2	1	0	3	ductal	le	4/12	1	0	pre	HE	5	2	4	0	1
26	47	2	0	0	2–3	ductal	le	0/1	1	1	pre	HE	0	0	3	0	2
27	29	2	0	0	2	mucinous	ri	0/14	1	1	pre	PET	2	1	4	0	1
28	52	2	1 bii	0	2	ductal	ri	21/36	1	1	post[7]	PET	0	0	0	0	2
29	35	2	0	0	2–3	ductal	ri	0/8	1	1	pre	PET	3	2	3	0	1
30	37	2	1 bii	1	3	ductal	ri	4/12	1	1	pre	PET,HE	4	2	8	1	1
31	42	4b	1	0	3	lobuloductal	le	17/17	1	1	pre	HE	2	1	4	0	1
32	35	2	1 biv	0	3	ductal	le	3/6	1	0	pre	PET	3	2	4	0	1
33	45	2	1 biv	x	3	ductal	ri	3/20	1	1	pre[7]	HE	1	1	1	0	2
34	35	2	1	1	2	ductal	ri	6/19	1	0	pre	PET	1	1	2	0	1
35	41	2	1 biv	1	3	ductal	le	11/20	0	0	pre	PET	1	1	2	0	0
36	45	2(m)	1 bi	0	2	lobular	le	3/16	0	0	pre	PET	3	2	2	0	1
37	44	1c	1 biii	1	2	ductal	le	2/22	1	0	post	PET	4	2	3	0	1
38	56	2	1	0	3	ductal	ri	5/14	1	1	post[7]	HE	1	1	1	0	2
39	34	1c	0	x	2	ductal	ri	x	(1)	1	pre	PET	3	2	1	0	1

Staging and grading (T, N, M, G) according to the UICC TNM Classification of Malignant Tumors [11].
PET = Paraffin-embedded tissue; FT = frozen tissue; TP = touch preparation; HE = HE-stained slide; x = data not available.

[1] Age at diagnosis.
[2] Estrogen receptor: 0 = negative; 1 = positive.
[3] Progesterone receptor: 0 = negative; 1 = positive.
[4] MSI ≥20%. MSI status: 0 = stable; 1 = low instable; 2 = high instable.
[5] LOH ≥40%. CIN status: 0 = stable; 1 = instable.
[6] Tumor heterogeneity: 0 = homogenous; 1 = heterogenous; 2 = not available.
[7] Limits for uncertain menopausal status: ≤50 years: premenopausal; >50 years: postmenopausal.

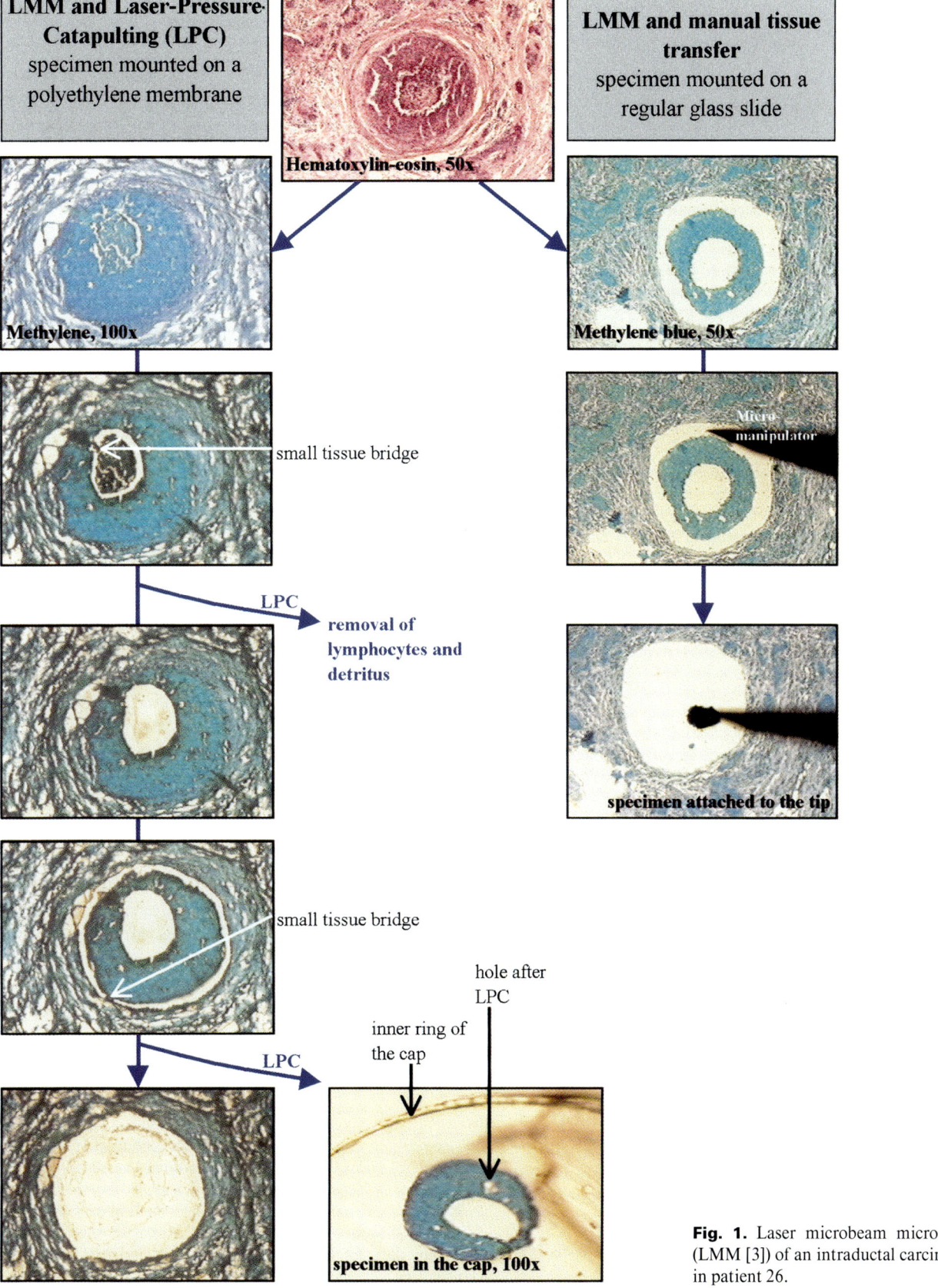

Fig. 1. Laser microbeam microdissection (LMM [3]) of an intraductal carcinoma area in patient 26.

Microsatellite Analyses

15 microsatellite markers on chromosomes 1, 2, 4, 5, 6, 11, 17, 18, 19 and 21 were analyzed. There is no established microsatellite primer panel for the detection of MSI in breast cancer. Therefore, the recommended reference panel for detection of MSI in colorectal cancer was used. This panel is composed of two mononucleotide repeats (BAT25, BAT26) and three dinucleotide repeats (D5S346, D2S123, D17S250) [15]. Additional markers with high sensitivity for MSI detection in colorectal cancer were included (BAT40, TP53Als [14, 16]. In addition, 8 tri- and tetranucleotide markers with the highest rate of MSI in breast cancer were selected according to the literature. Figure 2 shows the characteristics of the microsatellite markers analyzed [16–28].

PCR amplifications were performed with 50–100 ng of purified genomic DNA in a final volume of 20 μl in an MJ Research Thermocycler (PTC100, MJ Research, Watertown, Mass.) as described previously [29]. Subsequently, PCR products were analyzed by 6.7% polyacrylamide/50% urea gel electrophoresis (1 h, 1,500 V, 55°C) in a SequiGen sequencing gel chamber (BioRad, Hercules, Calif.) followed by silver nitrate staining [30].

Microsatellite Instability

MSI was defined by the presence of novel bands following PCR amplification of tumor DNA which were not present in PCR products of the corresponding normal DNA. All gels were evaluated independently by three different investigators (P.W., A.H. and W.D.). Every tumor was scored as MSI high, MSI low or microsatellite stable (MSS). A tumor was considered MSI high if 20% or more (3/15) of the examined loci showed instabilities, whereas less than 20% instability was classified as MSI low. MSS referred to no microsatellite locus being found to be unstable.

Chromosomal Instability

LOH referred to the signal intensity of a tumor allele relative to the matched normal allele being reduced by at least 50%. Chromosomal instability (CIN) was assumed if 40% (6/15) or more investigated markers revealed LOH. All tests were repeated at least twice in cases of MSI or LOH to exclude false-positive results.

Documentation

Regions of special interest of every case were captured with a digital camera (Hamamatsu color CCD camera C4200, Hamamatsu Photonics Deutschland, Herrsching, Germany) before microdissection and arranged for an overview (fig. 3) on a CorelDraw 4.0 surface, allowing an accurate assignment of genetic alterations to the histological phenotype.

Results

Microsatellite Instability

13 of 39 tumors (33%) showed MSI in at least 3 of the 15 investigated markers (MSI high, see table 1). 20 additional tumors (51%) revealed MSI at 1 or 2 microsatellite markers (MSI low). The remaining 6 tumors did not show MSI in any of the investigated markers. There were no differences in the clinical and histopathological features between MSI-positive and MSI-negative tumors.

Chromosomal Instability

9 of 39 tumors (23%) showed LOH at 6 or more investigated chromosomal regions. These specimens with frequent chromosomal deletions were defined as tumors with an instable chromosomal genotype (CIN-positive, see table 1). There was no correlation between MSI and CIN (p = 1.00) as previously reported for colorectal cancer [31]. The highest frequency of LOH was found for the chromosomal regions 17p13.1 (p53) and D17S125 (BRCA1) with 85 and 52% of the informative cases without MSI at this marker, respectively.

Value of Microdissection for Detection of Microsatellite Alterations

17 tumors were investigated in terms of different microdissection techniques using 15 microsatellite markers. Identical areas of these tumors were microdissected both manually and either by laser-mediated microdissection (n = 13) or by touch preparations of frozen tumor tissue (n = 4). 53 of the investigated 255 marker loci (21%) were not available for the comparison of manual and laser microdissection due to nonreproducible amplification results. Laser-based microdissection revealed LOH in 30 out of 202 markers (14.8%). In contrast, the manually microdissected samples showed LOH only in 17 out of 202 marker loci (8.4%). This difference was statistically significant (p = 0.043). 43% more LOH were detected in the samples in which laser microdissection or touch preparation was used in comparison to the manually dissected samples. In contrast, almost identical frequencies of MSI could be detected using manual or laser microdissection (13 vs. 14 MSI).

Tumor Heterogeneity

29 of 30 tumors, in which 2–6 different tumor regions were analyzed, showed genetic heterogeneity. This included both the deletion of different chromosomal loci in the separately processed specimens of one tumor, and a considerable difference in MSI within an MSI-positive tumor. Figure 4 shows representative examples of the genetic heterogeneity detected within different tumor regions.

Fig. 2. Microsatellite markers, primer sequences, PCR conditions and selected examples for MSI and LOH for each marker. Triangles show MSI or LOH. [a] Chrom. loc = Chromosomal localization; NA = not available. [b] Nonrepetitive nucleotides are indicated as dots. [c] PCR-Tm = PCR annealing temperature. [d] Examples for MSI and LOH; normal tissue is shown on the left, tumor on the right side. [e] Size according to: http://www.ncbi.nlm.nih.gov, http://www.gdb.org, http://www.resgen.com.

Repeat type	Name(s) (locus)	Chrom. Loc.[a]	Repeat motif[b]	Primer sequence (5' to 3')	PCR-Tm[c]	Size (bp)	Reference	MSI[d]	LOH[d]
Mononucleotide	BAT-25	4q12	TTTT T TTTT (T)₇ A(T)₂₅	TCG CCT CCA AGA ATG TAA GT / TCT GCA TTT TAA CTA TGG CTC	58°C	~90	Papadopoulos et al. [17]		
	BAT-26	2p16	(T)₅......(A)₂₆	TGA CTA CTT TTG ACT TCA GCC / AAC CAT TCA ACA TTT TTA ACC C	58°C	~80-100	Papadopoulos et al. [17]		
	BAT-40	1p13.1	TTTT TT...(T)₇......TTTT (T)₄₀	ATT AAC TTC CTA CAC CAC AAC / GTA GAG CAA GAC CAC CTT G	58°C	~80-100	Liu et al. [18]		
Pure dinucleotide	APC (D5S346)	5q21-5q22	(CA)₂₆	ACT CAC TCT AGT GAT AAA TCG / AGC AGA TAA GAC AGT ATT ACT AGT T	55°C	96-122	Spino et al [19]		
	1019/1020, MFD257 (D11S988)	11p15.5	(..)₁₆(CA)₂₅........	CAG AAA ATA GTT CAG ACC ACC A / GGG ACA AGA GAA AGT TGA ACA	58°C	~124*	Karnik et al. [20]		
Complex dinucleotide	Mfd15CA (D17S250)	17q11.2-q12	(TA)₅..........(CA)₂₄	GGA AGA ATC AAA TAG ACA AT / GCT GGC CAT ATA TAT ATT TAA ACC	52°C	~150	Weber et al. [21]		
	AFM093xb3 (D2S123)	2p16	CA₁₃TA(CA)₁₅(T/G A)₇	AAA CAG GAT GCC TGC CTT TA / GGA CTT TCC ACC TAT GGG AC	60°C	197-227	Weissenbach et al. [22]		
	TP53.Als (D17S513)	17p13.1	(..)₅₁(AAAAT)₈(A)₇GAAAA...	TCG AGG AGG TTG CAG TAA GCG GA / AAC AGC TCC TTT AAT GGC AG	60°C	~150*	Dietmaier et al. [16]		
Trinucleotide	DM	19q13.3	........(CTG)₁₁ GGGGG(..)₄₆	CTT CCC AGG CCT GCA GTT TGC CCA TC / GAA CGG GGC TCG AAG GGT CCT TGT AGC	67°C 130*, Elongation 72°C, 2'	~145*	Wooster et al. [23]	NA	
	TBP	6q27	(CAG)₃(CAAA)(CAG)₃CAACAGCAA(CAG)₁₈	CCC ACA GCC TAT TCA GAA CAC / GTT GAC TGC TGA ACG GCT GC	63°C	185-206	Polymenopoulos et al. [24]		
Tetranucleotide	Myc11	1p32	GAAAA(GAAA)₂TAAA(A/G)₁₀GAAAGA(GAAA)₁₄ GAAA(GAAAA)₂GAAAAA(GAAAA)₇	TGG CGA GAC TCC ATC AAA G / CTT TTT AAG CTG CAA CAA TTT C	53°C	140-209	Makela et al. [25]		
	UT574 (D18S51)	18q21.3	(..)₅₁(GAAAA)₃GAAN(GAAAA)₃...GAAAAA.. (GAAA)₂..GAAA(..)₁₃₄	GAG CCA TGT TCA TGC CAC TG / CAA ACC CGA CTA CCA GCA AC	58°C	~293*	Aldaz et al. [26]		
	GATA4H09 (D1S549)	1q32-q42	(..)₅₆CTAT........(CTAT)₁₁..........	CAA AGA GAA CAT GTG TTT GTG / TAC CAG CAA TGG GTA GTA TGG	58°C	~162*	Rosenberg et al. [27]		
	GGAA4D07 (D2S443)	2pter-2qter	A(AAGG)₁₈AAAAA(..)₁₃₂	GAG AGG GCA AGA CTT GGA AG / ATG GAA GAG CGT TCT AAA ACA	58°C	185-258	Paulson et al. [28]		
	GGAA2E02 (D21S1436)	21pter-21qter	(AAGG)₁₀GNG(..)₈₀	AGG AAA GAG AAA GAA AGG AAG G / TAT ATG ATG AAA GTA TAT TGG GGG	58°C	172-219	Paulson et al. [28]		

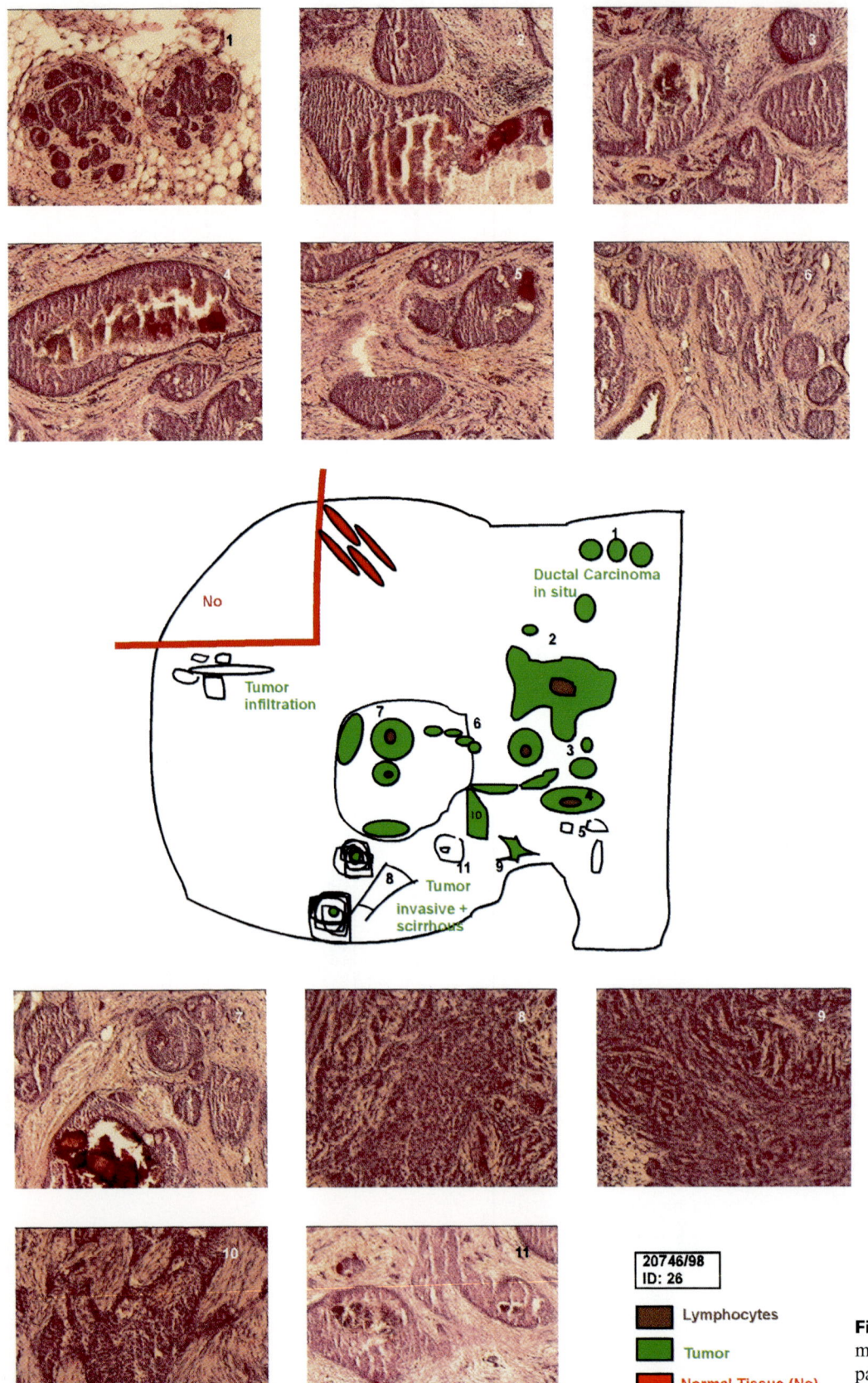

Fig. 3. Schematic illustration of microdissected tumor areas in patient 26 and captured tumor regions before microdissection.

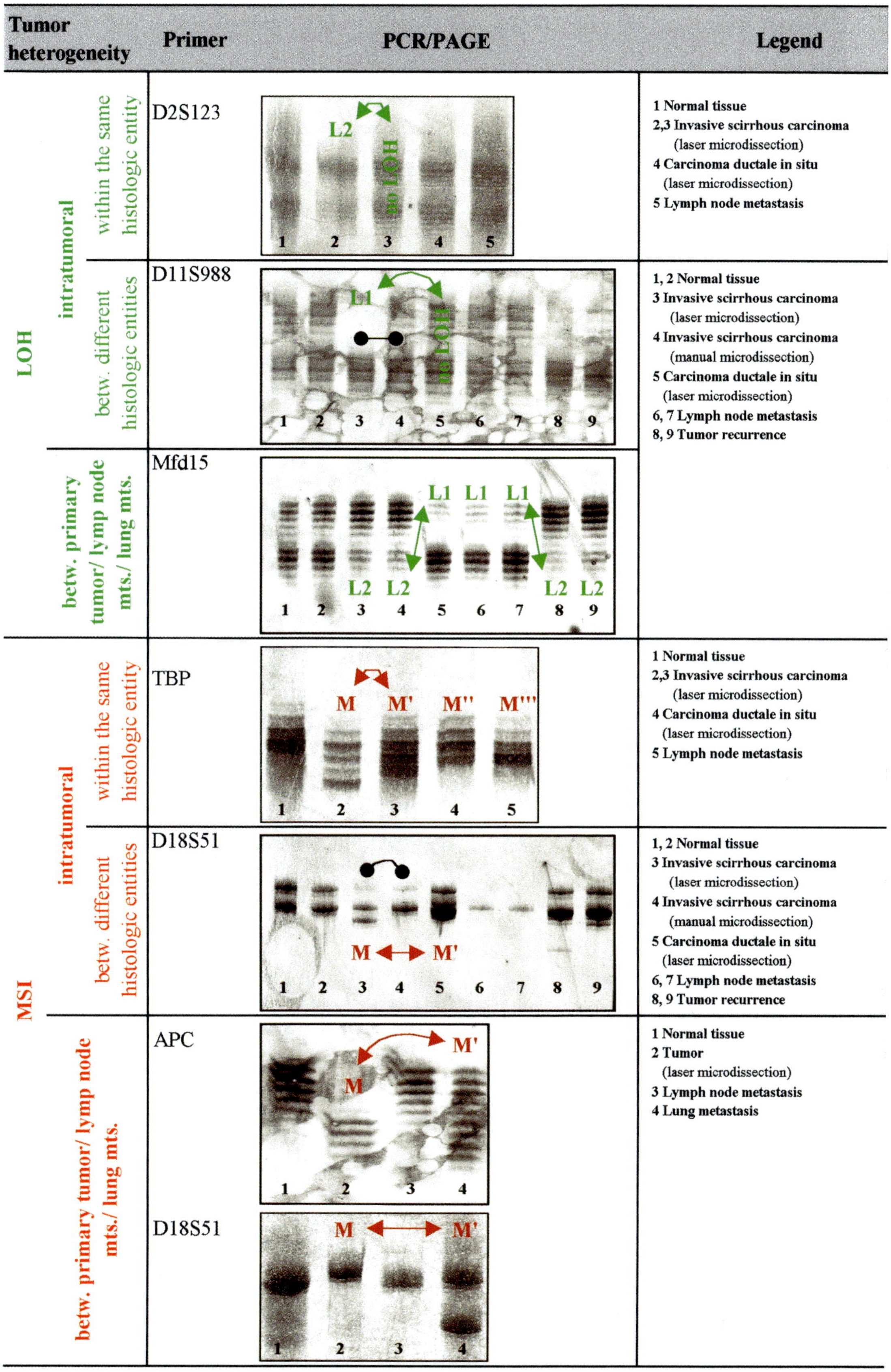

Tumor heterogeneity		Primer	PCR/PAGE	Legend
LOH	intratumoral — within the same histologic entity	D2S123		1 Normal tissue 2,3 Invasive scirrhous carcinoma (laser microdissection) 4 Carcinoma ductale in situ (laser microdissection) 5 Lymph node metastasis
	intratumoral — betw. different histologic entities	D11S988		1, 2 Normal tissue 3 Invasive scirrhous carcinoma (laser microdissection) 4 Invasive scirrhous carcinoma (manual microdissection) 5 Carcinoma ductale in situ (laser microdissection) 6, 7 Lymph node metastasis 8, 9 Tumor recurrence
	betw. primary tumor/ lymph node mts./ lung mts.	Mfd15		
MSI	intratumoral — within the same histologic entity	TBP		1 Normal tissue 2,3 Invasive scirrhous carcinoma (laser microdissection) 4 Carcinoma ductale in situ (laser microdissection) 5 Lymph node metastasis
	intratumoral — betw. different histologic entities	D18S51		1, 2 Normal tissue 3 Invasive scirrhous carcinoma (laser microdissection) 4 Invasive scirrhous carcinoma (manual microdissection) 5 Carcinoma ductale in situ (laser microdissection) 6, 7 Lymph node metastasis 8, 9 Tumor recurrence
	betw. primary tumor/ lymph node mts./ lung mts.	APC		1 Normal tissue 2 Tumor (laser microdissection) 3 Lymph node metastasis 4 Lung metastasis
		D18S51		

Fig. 4. Examples of tumor heterogeneity and differences in the detection of microsatellite alterations between manual and laser microdissected samples. PCR = Polymerase chain reaction; PAGE = polyacrylamide gel electrophoresis; mts. = metastasis; L1 = loss of longer allele; L2 = loss of shorter allele. Double arrows show tumor heterogeneity. Double dots show differences in the detection of microsatellite alterations between manual and laser microdissection.

In 11 tumors both intraductal and invasive areas of the tumor were investigated with 15 microsatellite markers. From an overall of 165 markers in these patients, 40 were not reproducibly amplified in both tumor areas and not available for comparison. In all 11 tumors, there was heterogeneity of the genetic alterations. Invasive tumor areas showed a higher percentage of genetic alterations than intraductal tumor areas both in LOH analyses (36/115; 31%) and MSI (31/115; 27%). In contrast, the intraductal components of the tumor had an LOH in only 20 of 115 markers (17%; p = 0.013) and MSI in 15 of 115 markers (13%; p = 0.008). Interestingly, in 3 of 5 patients with MSI-positive tumors only the invasive tumor component showed frequent MSI in 3 or more markers.

Discussion

The aim of this study was to evaluate the value of accurate microdissection for microsatellite analyses in breast cancer. Microsatellite analysis using highly polymorphic markers is frequently used for the detection of chromosomal deletions by LOH analysis. In the past, many of these studies used crude frozen tumor tissue or paraffin-embedded material without any microdissection of the tumor cells. In many tumor types this leads to the investigation of tumor DNA which is considerably diluted by normal DNA derived from stromal and inflammatory cells. It was clearly demonstrated previously that at least 70–80% tumor cells are needed for reliable detection of LOH [10]. In many of the LOH studies not using microdissection techniques, there is probably a considerable underestimation of the frequency of the detected chromosomal deletions.

In this study, the detection rate of LOH was 43% higher with laser microdissection in comparison to manual microdissection. Although manual microdissection is a fast, affordable and easy method for large homogenous tumor areas, it is very difficult to obtain a pure tumor cell population suited for LOH analyses, especially in tumor types with frequent productive fibrosis (scirrhous growth) as in breast cancer.

There was no difference in the detection rates of MSI between laser-assisted and manual microdissection. This is an expected result because in MSI additional alleles are detected. Therefore, this technique can be reliably used in samples with higher contamination with normal cells. However, our own results suggest that the detection rate of MSI decreases if the investigated sample contains less than 50% of tumor cells (data not shown). This could explain why the frequency of MSI detected in this study was higher as previously reported in the literature [23, 26, 28, 32–39]. MSI in at least two of the investigated markers was only detected in 0–30% of all samples. Taking all these studies together, there was MSI in 61 out of 867 breast cancers (7%). However, microdissection was used only in 2 of the 11 investigations. In the present study all samples were carefully microdissected, probably increasing the chance to detect MSI in breast cancer. Another explanation for the higher rate of MSI is that a highly selected marker set was used. Besides the recommended consensus primer panel for detection of MSI in colorectal cancer, markers which showed the highest MSI detection rates in previous studies were selected.

A second important result of this study is the demonstration of a considerable tumor heterogeneity. Almost all investigated breast cancer patients showed varying genetic alterations in different regions of the tumor. This was especially true for MSI-positive tumors which also showed an intratumoral MSI. The instability of repetitive sequences within one tumor is also an important feature of colorectal cancers or other tumors frequently occurring in patients with the human nonpolyposis colorectal cancer syndrom and is due to defects in DNA mismatch repair [15, 18].

In addition, heterogeneity in the LOH pattern between different tumor regions was found. The markers used in this study were not selected in order to detect regions frequently deleted in breast cancer, with the exception of D17S250 (BRCA1) and p53Als (p53). As expected, these two loci showed the highest frequency of LOH in 85% (p53) and 58% (BRCA1) of all informative cases. An interesting finding was the statistically significant higher frequency of both LOH and MSI in invasive carcinomas in comparison to the intraductal tumor in the same patient. The intraductal tumor component showed only 55% of the LOH and 48% of the MSI compared to the invasive tumor. Because we did not select specific chromosomal regions implicated in the progression of breast cancer, the data favor the hypothesis that genomic instability increases in general during the progression from intraductal to invasive carcinoma.

Another important result of the study is that in 3 patients MSI at 3 or more markers could be detected in the invasive tumor, but not in the intraductal carcinoma. This suggests that the MSI phenotype in breast cancer occurs later in the progression of the disease. This fact could provide another explanation for the high frequency of MSI found in this study in which only clinically advanced breast cancer patients treated with high-dose chemotherapy and autologous stem cell transplantation were included.

In conclusion, we could show that accurate microdissection is a valuable and necessary tool in molecular oncology and allows a precise determination of genetic alterations in histologically different tumor regions.

Acknowledgments

We thank Andrea Schneider and Monika Kerscher for excellent technical support in the areas of tissue processing and LOH analyses.

References

1 Sirivatanauksorn Y, Drury R, Crnogorac-Jurcevic T, Sirivatanauksorn V, Lemoine NR: Laser-assisted microdissection: Applications in molecular pathology. J Pathol 1999;189:150–154.
2 Suarez-Quian CA, Goldstein SR, Pohida T, Smith PD, Peterson JI, Wellner E, Ghany M, Bonner RF: Laser capture microdissection of single cells from complex tissues. Biotechniques 1999;26:328–335.
3 Schutze K, Lahr G: Identification of expressed genes by laser-mediated manipulation of single cells. Nat Biotechnol 1998;16:737–742.
4 Schutze K, Posl H, Lahr G: Laser micromanipulation systems as universal tools in cellular and molecular biology and in medicine. Cell Mol Biol (Noisy-le-Grand) 1998;44:735–746.
5 Becker I, Becker KF, Rohrl MH, Hofler H: Laser-assisted preparation of single cells from stained histological slides for gene analysis. Histochem Cell Biol 1997;108:447–451.
6 Bohm M, Wieland I, Schutze K, Rubben H: Microbeam MOMeNT: Non-contact laser microdissection of membrane-mounted native tissue. Am J Pathol 1997;151:63–67.
7 Schutze K, Becker I, Becker KF, Thalhammer S, Stark R, Heckl WM, Bohm M, Posl H: Cut out or poke in – The key to the world of single genes: Laser micromanipulation as a valuable tool on the look-out for the origin of disease. Genet Anal 1997;14:1–8.
8 Bonner RF, Emmert-Buck M, Cole K, Pohida T, Chuaqui R, Goldstein S, Liotta LA: Laser capture microdissection: Molecular analysis of tissue. Science 1997;278:1481–1483.
9 Emmert-Buck MR, Bonner RF, Smith PD, Chuaqui RF, Zhuang Z, Goldstein SR, Weiss RA, Liotta LA: Laser capture microdissection. Science 1996;274:998–1001.
10 Bohm M, Wieland I: Analysis of tumor-specific alterations in native specimens by PCR: How to procure the tumor cells. Int J Oncol 1997;15: 63–67.
11 Sobin LH, Wittekind C: TNM Classification of Malignant Tumors, ed 5. New York, Wiley, 1997.
12 Kovach JS, McGovern RM, Cassady JD, Swanson SK, Wold LE, Vogelstein B, Sommer SS: Direct sequencing from touch preparations of human carcinomas: Analysis of p53 mutations in breast carcinomas. J Natl Cancer Inst 1991;83:1004–1009.

13 Zhang L, Cui X, Schmitt K, Hubert R, Navidi W, Arnheim N: Whole genome amplification from a single cell: Implications for genetic analysis. Proc Natl Acad Sci USA 1992;89:5847–5851.
14 Dietmaier W, Hartmann A, Wallinger S, Heinmoller E, Kerner T, Endl E, Jauch KW, Hofstaedter F, Ruschoff J: Multiple mutation analyses in single tumor cells with improved whole genome amplification. Am J Pathol 1999;154: 83–95.
15 Boland CR, Thibodeau SN, Hamilton SR, Sidransky D, Eshleman JR, Burt RW, Meltzer SJ, Rodriguez-Bigas MA, Fodde R, Ranzani GN, Srivastava S: A National Cancer Institute Workshop on Microsatellite Instability for cancer detection and familial predisposition: Development of international criteria for the determination of microsatellite instability in colorectal cancer. Cancer Res 1998;58:5248–5257.
16 Dietmaier W, Wallinger S, Bocker T, Kullmann F, Fishel R, Ruschoff J: Diagnostic microsatellite instability: Definition and correlation with mismatch repair protein expression. Cancer Res 1997;57:4749–4756.
17 Papadopoulos N, Nicolaides NC, Liu B, Parsons R, Lengauer C, Palombo F, D'Arrigo A, Markowitz S, Willson JK, Kinzler KW: Mutations of GTBP in genetically unstable cells. Science 1995;268:1915–1917.
18 Liu B, Parsons R, Papadopoulos N, Nicolaides NC, Lynch HT, Watson P, Jass JR, Dunlop M, Wyllie A, Peltomaki P, de la Chapelle A, Hamilton SR, Vogelstein B, Kinzler KW: Analysis of mismatch repair genes in hereditary nonpolyposis colorectal cancer patients. Nat Med 1996;2:169–174.
19 Spirio L, Joslyn G, Nelson L, Leppert M, White R: A CA repeat 30-70 KB downstream from the adenomatous polyposis coli (APC) gene. Nucleic Acids Res 1991;19:6348.
20 Karnik P, Plummer S, Casey G, Myles J, Tubbs R, Crowe J, Williams BR: Microsatellite instability at a single locus (D11S988) on chromosome 11p15.5 as a late event in mammary tumorigenesis. Hum Mol Genet 1995;4:1889–1894.

21 Weber JL, Kwitek AE, May PE, Wallace MR, Collins FS, Ledbetter DH: Dinucleotide repeat polymorphisms at the D17S250 and D17S261 loci. Nucleic Acids Res 1990;18:4640.
22 Weissenbach J, Gyapay G, Dib C, Vignal A, Morissette J, Millasseau P, Vaysseix G, Lathrop M: A second-generation linkage map of the human genome. Nature 1992;359:794–801.
23 Wooster R, Cleton-Jansen AM, Collins N, Mangion J, Cornelis RS, Cooper CS, Gusterson BA, Ponder BA, von Deimling A, Wiestler OD: Instability of short tandem repeats (microsatellites) in human cancers. Nat Genet 1994;6: 152–156.
24 Polymeropoulos MH, Rath DS, Xiao H, Merril CR: Trinucleotide repeat polymorphism at the human transcription factor IID gene. Nucleic Acids Res 1991;19:4307.
25 Makela TP, Hellsten E, Vesa J, Alitalo K, Peltonen L: An Alu variable polyA repeat polymorphism upstream of L-myc at 1p32. Hum Mol Genet 1992;1:217.
26 Aldaz CM, Chen T, Sahin A, Cunningham J, Bondy M: Comparative allelotype of in situ and invasive human breast cancer: High frequency of microsatellite instability in lobular breast carcinomas. Cancer Res 1995;55:3976–3981.
27 Rosenberg CL, De Las M, Huang K, Cupples LA, Faller DV, Larson PS: Detection of monoclonal microsatellite alterations in atypical breast hyperplasia. J Clin Invest 1996;98: 1095–1100.
28 Paulson TG, Wright FA, Parker BA, Russack V, Wahl GM: Microsatellite instability correlates with reduced survival and poor disease prognosis in breast cancer. Cancer Res 1996; 56:4021–4026.
29 Hartmann A, Rosner U, Schlake G, Dietmaier W, Zaak D, Hofstaedter F, Knuechel R: Clonality and genetic divergence in multifocal low-grade superficial urothelial carcinomas as determined by chromosomes 9 and p53 deletion analysis. Lab Invest 2000;80:709–718.
30 Schlegel J, Bocker T, Zirngibl H, Hofstaedter F, Ruschoff J: Detection of microsatellite instability in human colorectal carcinomas using a non-radioactive PCR-based screening technique. Virchows Arch 1995;426:223–227.

31 Lengauer C, Kinzler KW, Vogelstein B: Genetic instability in colorectal cancers. Nature 1997;386:623–627.
32 Yee CJ, Roodi N, Verrier CS, Parl FF: Microsatellite instability and loss of heterozygosity in breast cancer. Cancer Res 1994;54:1641–1644.
33 Contegiacomo A, Palmirotta R, De Marchis L, Pizzi C, Mastranzo P, Delrio P, Petrella G, Figliolini M, Bianco AR, Frati L: Microsatellite instability and pathological aspects of breast cancer. Int J Cancer 1995;64:264–268.
34 Shaw JA, Walsh T, Chappell SA, Carey N, Johnson K, Walker RA: Microsatellite instability in early sporadic breast cancer. Br J Cancer 1996;73:1393–1397.
35 Toyama T, Iwase H, Iwata H, Hara Y, Omoto Y, Suchi M, Kato T, Nakamura T, Kobayashi S: Microsatellite instability in in situ and invasive sporadic breast cancers of Japanese women. Cancer Lett 1996;108:205–209.
36 Dillon EK, de Boer WB, Papadimitriou JM, Turbett GR: Microsatellite instability and loss of heterozygosity in mammary carcinoma and its probable precursors. Br J Cancer 1997;76:156–162.
37 Rush EB, Calvano JE, van Zee KJ, Zelenetz AD, Borgen PI: Microsatellite instability in breast cancer. Ann Surg Oncol 1997;4:310–315.
38 Anbazhagan R, Fujii H, Gabrielson E: Microsatellite instability is uncommon in breast cancer. Clin Cancer Res 1999;5:839–844.
39 Richard SM, Bailliet G, Paez GL, Bianchi MS, Peltomaki P, Bianchi NO: Nuclear and mitochondrial genome instability in human breast cancer. Cancer Res 2000;60:4231–4237.

Cell Type-Specific mRNA Quantitation in Non-Neoplastic Tissues after Laser-Assisted Cell Picking

Rainer M. Bohle[a]   Elena Hartmann[a]   Thomas Kinfe[a]   Leander Ermert[a]
Werner Seeger[b]   Ludger Fink[a,b]

[a]Department of Pathology and [b]Internal Medicine, Justus Liebig University Giessen, Germany

Key Words

Laser-assisted cell picking · RT-PCR, real time · mRNA quantitation · Nitric oxide synthase II mRNA · Endotoxin priming · Lipopolysaccharide · Microdissection

Abstract

Objective: Cell type-specific mRNA quantitation can be reliably performed after harvesting less than 20 cell profiles from haemalaun-stained cryosections by laser-assisted cell picking. Up to now it has been unclear to what extent these techniques can be used to analyze differential gene expression in complex tissues. *Methods:* Using a rat model of experimental endotoxin priming of the lung various pulmonary cell types were microdissected from isolated perfused and ventilated rat lungs after aerosol lipopolysaccharide/interferon-γ stimulation. *Results:* Porphobilinogen deaminase housekeeping gene (PBGD) and nitric oxide synthase II (NOSII) mRNA in arterial endothelial cells (AEC), bronchiolar epithelial cells (BEC), alveolar septum containing monocytes/macrophages (AS+), alveolar septum without monocytes/macrophages (AS–) and intraluminar alveolar macrophages (AM) could be quantified by real-time RT-PCR. The strongest upregulation of NOSII mRNA occurred in AM, while minimal NOSII expression was detected in BEC, AS+ and AS–. In AEC NOSII mRNA was not detectable. *Conclusion:* The combination of laser microdissection and real-time RT-PCR is a valuable tool for the quantitative in situ characterization of differential gene expression within complex tissues.

Copyright © 2001 S. Karger AG, Basel

Introduction

Gene expression in non-neoplastic tissues or cell cultures is currently mainly analyzed by Northern blots, RNase protection assays or real-time RT-PCR after homogenization of cells and tissues. These methods allow only rough estimations of physiologically relevant mRNA alterations, as only average expression levels can be determined. Morphological techniques like in situ hybridization are more precise with respect to cell type-specific characterization of gene expression, but changes in expression levels can only be described semiquantitatively. Moreover, the most sensitive in situ labelling techniques for mRNA analysis in non-neoplastic tissues are time-consuming requiring radioactive labelling and up to several weeks for incubation. Microdissection combined with few cell real-time RT-PCR overcomes many of these methodological disadvantages. It allows cell type-specific mRNA isolation without contamination of adjacent cells and valid mRNA quantitation from complex tissues [1, 2]. Up to now it has not been clear whether this approach can also be used to characterize simultaneously the differential gene expression in various cell types from the same organ. For this purpose, we investigated the quantitative expression of nitric oxide synthase II (NOSII) in various cell types from rat lungs undergoing endotoxin priming with lipopolysaccharide (LPS) and interferon-γ (IFN-γ) nebulization, using UV laser-assisted cell picking (LACEP) for cell isolation.

Material and Methods

Experimental Model

Male CD rats (Sprague-Dawley, 60–70 days old, 350–400 g body weight; Charles River, Sulzfeld, Germany) were deeply anaethetized with phenobarbital-sodium (100 mg/kg body weight i.p.). Lungs were carefully isolated, ventilated and perfused as described previously [3, 4]. For stimulation, an ultrasonic nebulizer was employed to aerosolize 75 μg LPS and 1,000 U IFN-γ in a total volume of 5 ml into the afferent limb of the ventilator circuit within 10 min. Experiments were then continued under standard conditions for 6 h. For fixation, lungs were perfused with 4.5% formaldehyde solution (Roti-Histofix, Roth, Karlsruhe, Germany) for 15 min, followed by a rinsing step with Krebs-Henseleit buffer to remove the residual formalin. Finally, right lung was instilled with Tissue Tek (Sakura Finetek, Zoeterwoude, The Netherlands) and snap-frozen in liquid nitrogen.

LACEP, mRNA Isolation and cDNA Synthesis

Cell picking was performed as described in detail recently [1, 5]. Using the UV laser Robot Microbeam (PALM, Bernried, Germany), arterial endothelial cells (AEC), bronchiolar epithelial cells (BEC), alveolar septum containing monocytes/macrophages (AS+), alveolar septum without monocytes/macrophages (AS–) and intraluminal alveolar macrophages (AM) were selected microscopically and harvested after laser microdissection. Clearly recognizable monocytes/macrophages from the alveolar septum (AS–) were removed by UV laser-mediated photolysis [6] during brightfield illumination. Adjacent tissue was removed by UV laser photolysis under visual control. Finally, the cell/alveolar septum profiles were isolated by sterile syringe needles and transferred into a reaction tube containing 10 μl first-strand buffer. Samples with 20 cell profiles each were snap-frozen in liquid nitrogen and stored for further preparation. After a short thawing period, proteinase K (1 μl; 1 mg/ml; Sigma, Deisenhofen, Germany) was added to the samples and incubated for 1 h at 53°C. Proteinase K as well as RNA were denaturated for 5 min at 95°C, and samples were stored on ice for another 5 min. cDNA synthesis was performed using 2 μl MgCl$_2$ (25 mM), 2 μl GeneAmp 10 × PCR Buffer II (100 mM Tris-HCl, pH 8.3, 500 mM KCl), 1 μl dNTP (10 mM each), 1 μl random hexamers (50 μM), 0.5 μl RNase inhibitor (10 U) and 1 μl MuLV reverse transcriptase (50 U). Except for dNTP (Eurobio, Raunheim, Germany), all reagents were purchased from Applied Biosystems (Weiterstadt, Germany). Samples were incubated at 20°C for 10 min and at 43°C for 60 min. Finally, the reaction was stopped by 5 min at 99°C.

Relative mRNA Quantitation

Real-time PCR was used for relative quantitation of the porphobilinogen deaminase housekeeping gene (PBGD) low copy housekeeping gene mRNA [2] and the NOSII mRNA. It is based on the 5′ nuclease activity of *Taq* polymerase for fragmentation of a dual-labeled fluorogenic hybridization probe [7, 8]. Using the ABI Prism 7700 Sequence Detection System (Applied Biosystems), it was performed as recently described in detail [1]. For quantitation, the target gene was normalized to the internal standard gene, PBGD. This kind of quantitation is calculated by the following equation:

$$\frac{T_o}{R_o} = K \cdot (1 + E)^{(CT,R - CT,T)}$$

where T_o = initial number of target gene mRNA copies, R_o = initial number of standard gene mRNA copies, E = efficiency of amplification, CT,T = threshold cycle of target gene, CT,R = threshold cycle of standard gene and K = constant.

The following primer/probe sets showed in pilot experiments that amplification efficiencies of PBGD and NOSII mRNA were approximately equal and amounted to 0.95 ± 0.02. K is assumed to be equal within a definite fluorogenic-labeled primer/probe system and thus does not influence the comparison of calculated relative ratios.

Primers:
PBGD forward 5′CAAGGTTTTCAGCATCGCTACCA3′ (exon 4)
PBGD reverse 5′ATGTCCGGTAACGGCGGC3′ (exon 1)
NOSII forward 5′CTCGCTGCATCGGCAG GAT3′ (exon 6)
NOSII reverse 5′AGTCATGCTTCCCATCGCTCC3′ (exon 8)
Probes:
PBGD hybridization probe
5′CCAGCTGACTCTTCCGGGTGCCCAC3′ (exon 4 to exon 3)
NOSII hybridization probe
5′CCTGCAGGTCTTCGATGCCCGGA3′ (exon 6 to exon 7)

After reverse transcription, each sample was divided into two aliquots of 8 μl for target gene and standard gene analysis. Twenty-five microlitres Universal Master Mix (Applied Biosystems), oligonucleotide primers (final concentration 900 nM) and hybridization probe (final concentration 200 nM) were added to an end volume of 50 μl. Cycling conditions were 95°C for 6 min, followed by 55 cycles of 95°C for 20 s, 61°C for 30 s and 73°C for 30 s.

Statistical Analysis

Analysis of variance followed by the Kruskal-Wallis test and Mann-Whitney test for unpaired data was used to evaluate differences among the groups. A value of p < 0.05 was considered significant. All data are means ± SEM.

Results

Laser-Assisted Cell Picking

Target cells could be easily identified in the haemalaun-stained frozen sections, which were kept in 100% ethanol until microdissection. After focussing the laser

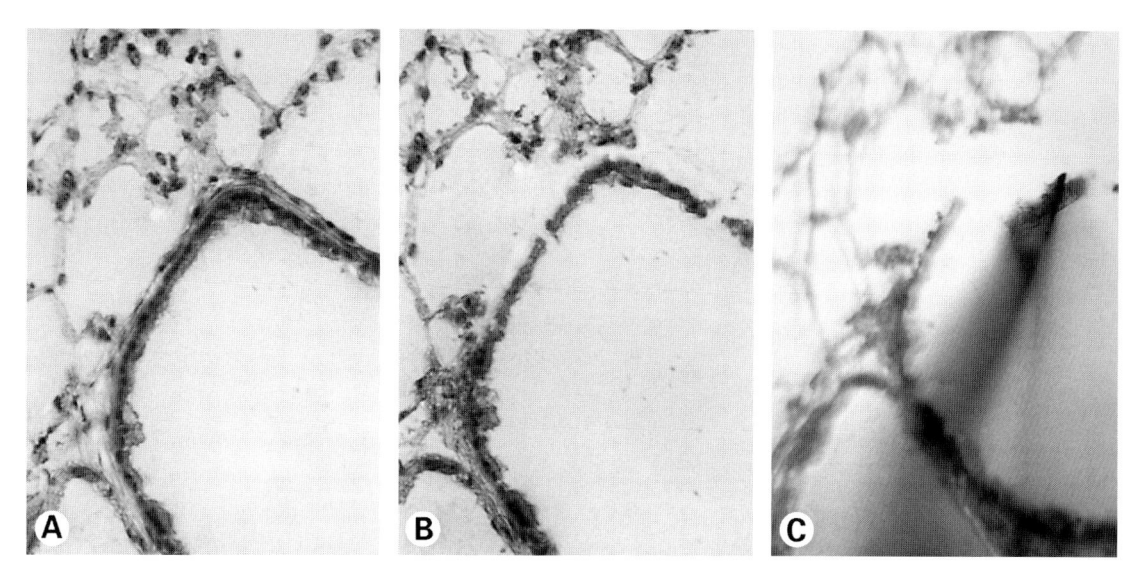

Fig. 1. LACEP in frozen rat lung tissue after haemalaun staining. **A** A bronchiolus is selected. **B** Microdissection of BEC by the UV laser. **C** The cell cluster is harvested with a needle and adheres tightly to the tip of the needle.

beam and adjusting the laser energy to a minimum the cells of interest were dissected from adjacent cells at 400× magnification. The precision of microdissection was checked by covering the unmounted haemalaun-stained tissue section with two drops of 100% ethanol, which improved histo- and cytomorphology considerably. A few seconds after evaporation, the target cells were isolated from the glass slide by adhesion to the tip of a sterile needle (30 gauge) (fig. 1), and were lifted by the computer-controlled micromanipulator for cell transfer into the reaction tube. Sterile needle tips dipped into the overlaying ethanol served as negative controls.

Efficiency Rates of PBGD and NOSII mRNA in Formalin-Fixed Stimulated Lung Tissue

Within formalin-fixed and haemalaun-stained tissue sections of the stimulated lungs, PBGD mRNA was detected in 51% of the samples containing 15–20 cell profiles or cell clusters with 15–20 nuclei. NOSII mRNA was found in 31.3% of the AS+ samples, 17.6% of the AS– samples, 7.1% of the BEC samples and in none of the AEC samples. Highest NOSII mRNA recovery was obtained for AM (38%) [9]. None of the negative controls were PBGD- or NOSII mRNA-positive.

Relative mRNA Quantitation

As a basis for the relative mRNA quantitation, mean threshold cycle for PBGD mRNA (CT_{PBGD}) was calculated. It amounted to 38.5 ± 0.24 (± SEM) for the stimulated lung. To achieve normalization of the non-regulated standard gene, the corresponding threshold cycles of the target gene samples were subtracted from mean PBGD CT ($\Delta CT = CT_{PBGD} - CT_{target\ gene}$; given as mean ± SEM). The ΔCT values for AM, AS+, AS–, BEC and AEC were –0.9 ± 0.4, –4.7 ± 0.4, –5.6 ± 0.5, –0.3 ± 0.3 and 0. Calculating a PCR efficiency of 0.95 ± 0.02 for the two investigated genes, the values of relative mRNA expression were obtained by the formula $K \cdot 1.95^{\Delta CT}$ (as depicted in Material and Methods). These values are given in figure 2.

Relative NOSII mRNA Expression in Laser-Microdissected Pulmonary Cells

Relative NOSII mRNA expression in picked AM, AS+, AS–, BEC and AEC amounted to 0.53 ± 0.17, 0.02 ± 0.01, 0.009 ± 0.006, 0.05 ± 0.05 and 0 copies/copy PBGD. The relative NOSII mRNA expression in the PBGD-positive samples was significantly higher in AM as compared to AS+ ($p < 0.02$), AS– ($p < 0.01$), BEC ($p < 0.04$) and AEC ($p < 0.01$). Furthermore, a significant accumulation of NOSII mRNA was noted in AS+ as compared to NOSII mRNA in AEC ($p < 0.04$). The relative NOSII mRNA expression levels did not differ significantly between samples from AS–, BEC and AEC.

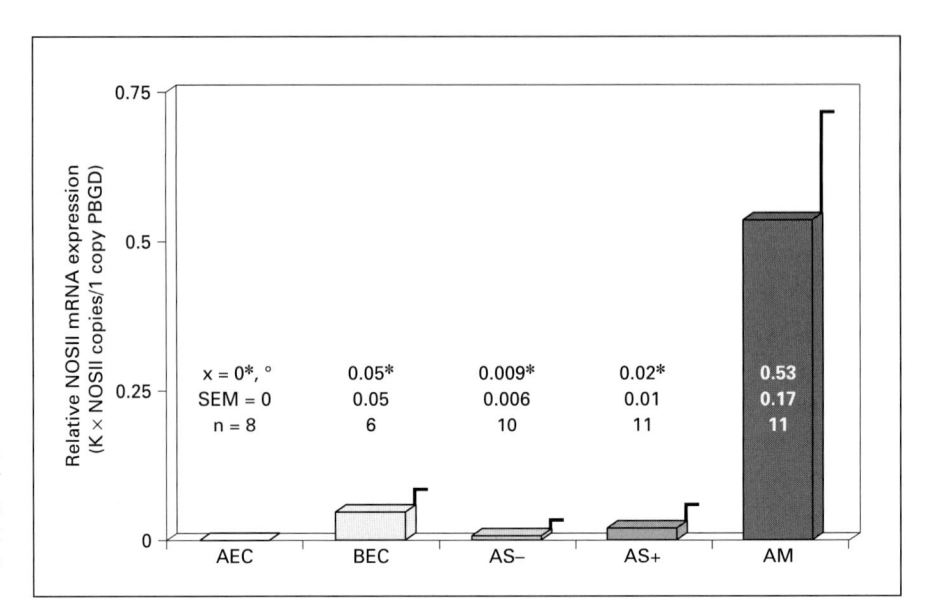

Fig. 2. Relative NOSII mRNA expression in AEC, BEC, AS–, AS+, and pure AM after LPS/IFN-γ nebulization. Data are given as mean SEM. K = Constant; * = significant compared to AM;° = significant compared to AS+.

Discussion

Pathophysiological events in non-neoplastic diseases are mainly dependent on humoral, microenvironmental or neural stimuli contributing to functional abnormalities. Exact measurements of various physiological parameters indicate that in complex organs, different cell types may be differently affected. For example, in the lungs exposed to endotoxin, vascular and bronchial cells seem to be the main target cells for TNF-α upregulation [10], while other cell types of the lung are probably less affected.

When aiming to characterize the cellular events on the mRNA level, most of the currently applied molecular techniques (e.g. Northern blots, RNase protection assays, mRNA quantitation methods) give only average expression levels, as they use homogenized tissue fragments. We recently described a fast and efficient technology to investigate cell type-specific gene expression in intact tissues in small numbers of isotypic cells [1]. Combining the precision of LACEP with the real-time RT-PCR technology and using minute amounts of mRNA, we showed that transcriptional regulation can be reliably quantified in defined cell populations.

To evaluate whether this technique can be applied to characterize differential gene expression on a quantitative basis in small tissue areas from complex organs, we investigated NOSII mRNA expression in closely neighbouring AM, alveolar septum cells with a few adjacent AM, alveolar septum cells without adjacent AM, isolated BEC and AEC. The well-established experimental model of ventilation and perfusion of isolate rat lungs [3, 4] served to stimulate pulmonary cells ex vivo by nebulization of endotoxin and IFN-γ. Prior to this study, quantitative gene expression after laser microdissection of less than 20 cell profiles was only investigated in AM and vascular endothelial cells originating from small intrapulmonary muscular vessels [1]. In that study TNF-α response to inflammatory challenge was observed in AM but not in vascular endothelial cells. Nevertheless, the questions remained open whether the LACEP procedure can be recommended for the investigation of neighbouring cells, as mRNA contamination might already appear during the process of cryosectioning.

Our results indicate that contamination due to RNA grease from adjacent cells is negligible. Quantitatively determined NOSII mRNA expression levels differed significantly between pure AM samples as compared with alveolar septum cells (AS+ and AS– samples). Moreover, significantly higher NOSII mRNA expression levels were detected in AS+ when compared to AS– samples, indicating the effect of residual AM on the expression level. In contrast, the very low levels of NOSII mRNA expression in AEC clearly suggest that false-positive expression (due to RNA contamination from adjacent cells) may not be relevant. Investigating sets of 1,000 laser capture microdissected rat neurons, Luo et al. [11] forwarded similar results, as they were able to detect significant variations of gene expression when neighbouring small and large neuron cells were procured.

In conclusion, the combination of laser microdissection and real-time RT-PCR allows quantitative and cell type-specific detection of differential gene expression in complex tissues. It is thus a valuable tool for precise in situ characterization of mRNA regulation in closely related non-neoplastic cells.

Acknowledgments

We thank M.M. Stein and G. Jurat for skillful technical assistance. This study was supported by the Deutsche Forschungsgemeinschaft (SFB 547 Cardiopulmonary Vascular System, project Z1).

References

1 Fink L, Seeger W, Ermert L, Hänze J, Stahl U, Grimminger F, Kummer W, Bohle RM: Real-time quantitative RT-PCR after laser-assisted cell picking. Nat Med 1998;4:1329–1333.

2 Fink L, Stahl U, Ermert L, Kummer W, Seeger W, Bohle RM: Rat porphobilinogen deaminase: A pseudogene-free internal standard for laser-assisted cell picking. Biotechniques 1999; 26:510–516.

3 Ermert L, Ermert M, Althoff A, Merkle M, Grimminger F, Seeger W: Vasoregulatory prostanoid generation proceeds via cyclo-oxygenase-2 in non-inflamed rat lungs. J Pharmacol Exp Ther 1998;286:1309–1314.

4 Seeger W, Walmrath D, Grimminger F, Rosseau S, Schutte H, Kramer HJ, Ermert L, Kiss L: Adult respiratory distress syndrome: Model systems using isolated perfused rabbit lungs. Methods Enzymol 1994;233:549–584.

5 Fink L, Kinfe T, Stein MM, Ermert L, Hänze J, Kummer W, Seeger W, Bohle RM: Immunostaining and cell picking for mRNA analysis. Lab Invest 2000; 80:327–333.

6 Kummer W, Fink L, Dvorakova M, Haberberger R, Bohle RM: Rat cardiac neurons express the non-coding R-exon (exon 1) of the cholinergic gene locus. Neuroreport 1998;9:2209–2212.

7 Heid CA, Stevens J, Livak KJ, Williams PM: Real time quantitative PCR. Genome Res 1996;6:986–994.

8 Gibson UEM, Heid CA, Williams PM: A novel method for real time quantitative RT-PCR. Genome Res 1996;6:995–1001.

9 Fink L, Kinfe T, Seeger W, Ermert L, Kummer W, Bohle RM: Immunostaining for cell picking and real-time mRNA quantitation. Am J Pathol 2000;157:1459–1466.

10 Ermert M, Merkle M, Mootz R, Grimminger F, Seeger W, Ermert L: Endotoxin priming of the cyclooxygenase-2-thromboxane axis in isolated rat lungs. Am J Physiol 2000;278:L1195-L1203.

11 Luo L, Salunga RC, Guo H, Bittner A, Joy KC, Galindo JE, Xiao H, Rogers KE, Wan JS, Jackson MR, Erlander MG: Gene expression profiles of laser-captured adjacent neuronal subtypes. Nat Med 1999;5:117–122.

Laser-Assisted Microdissection and Short Tandem Repeat PCR for the Investigation of Graft Chimerism after Solid Organ Transplantation

Wolfram Kleeberger[a] Thomas Rothämel[b] Sabine Glöckner[a]
Ulrich Lehmann[a] Hans Kreipe[a]

[a]Institute of Pathology and [b]Institute of Legal Medicine, Medizinische Hochschule Hannover, Germany

Key Words
Laser microdissection · Short tandem repeat PCR · Microchimerism · Endothelial cells · Hepatocytes

Abstract
The detection of donor-derived cells in the blood and tissues of graft recipients after solid organ transplantation is a readily observed phenomenon called microchimerism. Yet very little is known about the persistence and integration of recipient-derived cells in the transplanted organ, indicating a form of intragraft chimerism. To further study this phenomenon and its possible influence on graft acceptance or rejection, we developed the following novel approach. Immunohistochemically labeled cells were isolated by means of laser-based microdissection and subsequent laser pressure catapulting from paraffine-embedded posttransplantation biopsies. The following use of a highly sensitive PCR assay analyzing one polymorphic short tandem repeat (STR) marker enabled us to clearly identify the genotypes in samples containing as little as 10 isolated cells. The combination of laser-based microdissection and STR-PCR thus provides a powerful tool for the genotyping of even very few cells isolated from routinely processed biopsies after solid organ transplantation.

Copyright © 2001 S. Karger AG, Basel

Introduction

After transplantation of solid organs such as liver, heart, kidney and lungs, a small amount of donor-derived immunocompetent cells, mainly lymphocytes, are able to leave the graft and survive in the blood and in some peripheral organs of the recipient for years [1]. This phenomenon is commonly called microchimerism or more precisely, peripheral microchimerism and has been extensively studied during the last years [2–4]. Yet it still remains unclear whether it influences the recipient's immune response to the graft of whether it is just an epiphenomenon occurring after transplantation without considerable influence on the graft's fate [5].

On the other hand, only very little is known about the possibility that few cells of recipient origin enter the graft and are able to persist there permanently, indicating a form of intragraft microchimerism [6] (fig. 1). Support for the existence of such an intragraft chimerism comes from studies mainly on animal models. After creating suitable experimental conditions with combined liver and bone marrow transplantations it was possible to observe that after chemically induced liver damage the regeneration of hepatic and ductular cells was at least partially done by circulating precursor cells of bone marrow origin [7]. Moreover, using in situ hybridization techniques, recent studies found that hepatocytes and bile duct cells are partially substituted by such bone marrow-derived precursor cells also in humans after solid organ transplantation [8, 9].

In addition in vitro studies were able to show that circulating stem cells also can differentiate into adult endothelial cells [10, 11] or even turn into skeletal muscle [12]. However, there are only some recent studies searching for an intragraft chimerism on the endothelial level, all of which examined only very few cases and none of which could detect endothelial cells of recipient origin [13, 14].

The aim of this study, therefore, was to systematically examine in a larger number of cases whether and to what extent hepatocytes and endothelial cells or recipient origin can be integrated into grafts and to correlate the results with histopathological findings and other relevant parameters such as time interval between transplantation and biopsy taking.

To overcome the limitations and problems of the previously used in situ hybridization methodologies, we developed a new technical approach combining laser-assisted microdissection with subsequent PCR of highly polymorphic short tandem repeat (STR) markers in order to identify whether the sampled cells are of donor or of recipient origin.

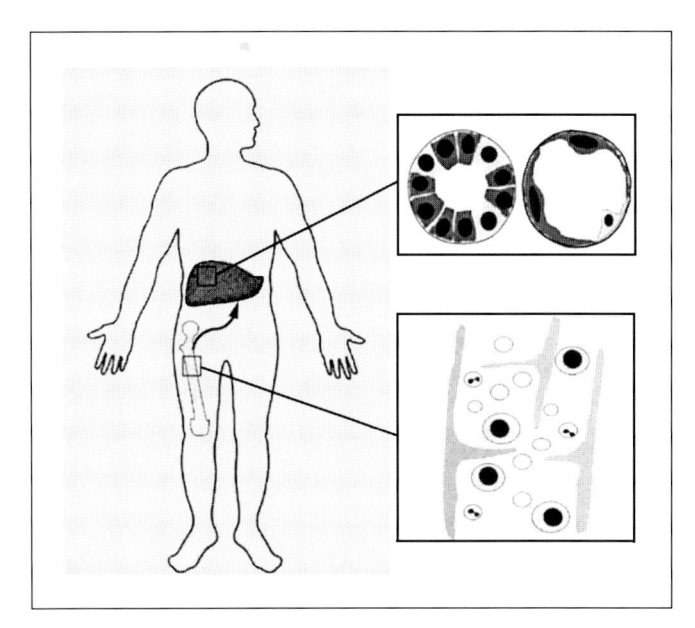

Fig. 1. Intragraft microchimerism. After liver transplantation recipient-derived precursor cells of bone marrow origin (light cells) enter the graft (dark cells) and become integrated as adult epithelial or endothelial cells.

Material and Methods

Material

Exclusively formalin-fixed, paraffin-embedded posttransplantation biopsies of livers and hearts were used for this study, all of them retrieved from the surgical pathology files of the Institute of Pathology of the Hannover Medical School (n = 23).

Slide Preparation and Immunohistochemical Staining

The slides were prepared as follows. First, an ultrathin polyethylene foil was fixed on the glass slides. 3- to 5-µm-thick paraffin sections were then mounted onto this foil. After deparaffinizing the tissue following standard protocols an immunohistochemical staining was performed in order to either identify the target cells themselves or to mark surrounding and thus possibly contaminating circulating blood cells. Immunostaining was performed according to the manufacturer's recommendations (BioGenex Super Sensitive Kit). Endothelial cells were labeled with CD31 (DAKO), lymphocytes and monocytes were identified using a combination of CD45 and CD68 (both DAKO). Finally the slides were counterstained with hematoxylin.

Laser Microdissection

Laser-based microdissection was performed using the PALM Laser MicroBeam system, essentially as described earlier [15], in order to collect the cells of interest (fig. 2). For isolation of the DNA 10 µl of proteinase K digestion buffer (50 mmol/l Tris; pH 8.1; 1 mmol/l EDTA; 0.5% Tween 20; 40 µg/ml proteinase K) were put into the lid which was then incubated overnight at 40°C. After denaturation at 95°C for 8 min, the whole lysate was used for subsequent PCR analysis.

Short Tandem Repeat PCR

To determine whether the sampled cells are of donor or of recipient origin a PCR assay analyzing highly polymorphic STR markers was developed. In contrast to previous studies using commercially available multiplex PCR kits [16–18], we chose one tetranucleotide repeat marker with a remarkably high heterozygosity rate of 93%, commonly known as SE 33 [19, 20].

PCR was performed in a final reaction volume of 25 µl containing 10× reaction buffer with 1.5 mM MgCl$_2$, 250 nmol/l of each dNTP, 400 mg/l BSA, 150 nmol/l of each primer and 0.5 U of Quiagen Hot Star Taq Polymerase. First the reaction mixture was preheated at 95°C for 15 min, followed by 30 cycles at 95°C for 30 s, 56°C for 30 s and 70°C for 1 min with a final elongation step of 70°C for 10 min.

PCR products were visualized using an ABI-PRISM capillary sequencer.

Results

Evaluation of Sensitivity

In order to test linearity, reproducibility and sensitivity of the method and in order to find out how many cells are needed to get reliable results we performed experiments in which defined cell counts were catapulted into

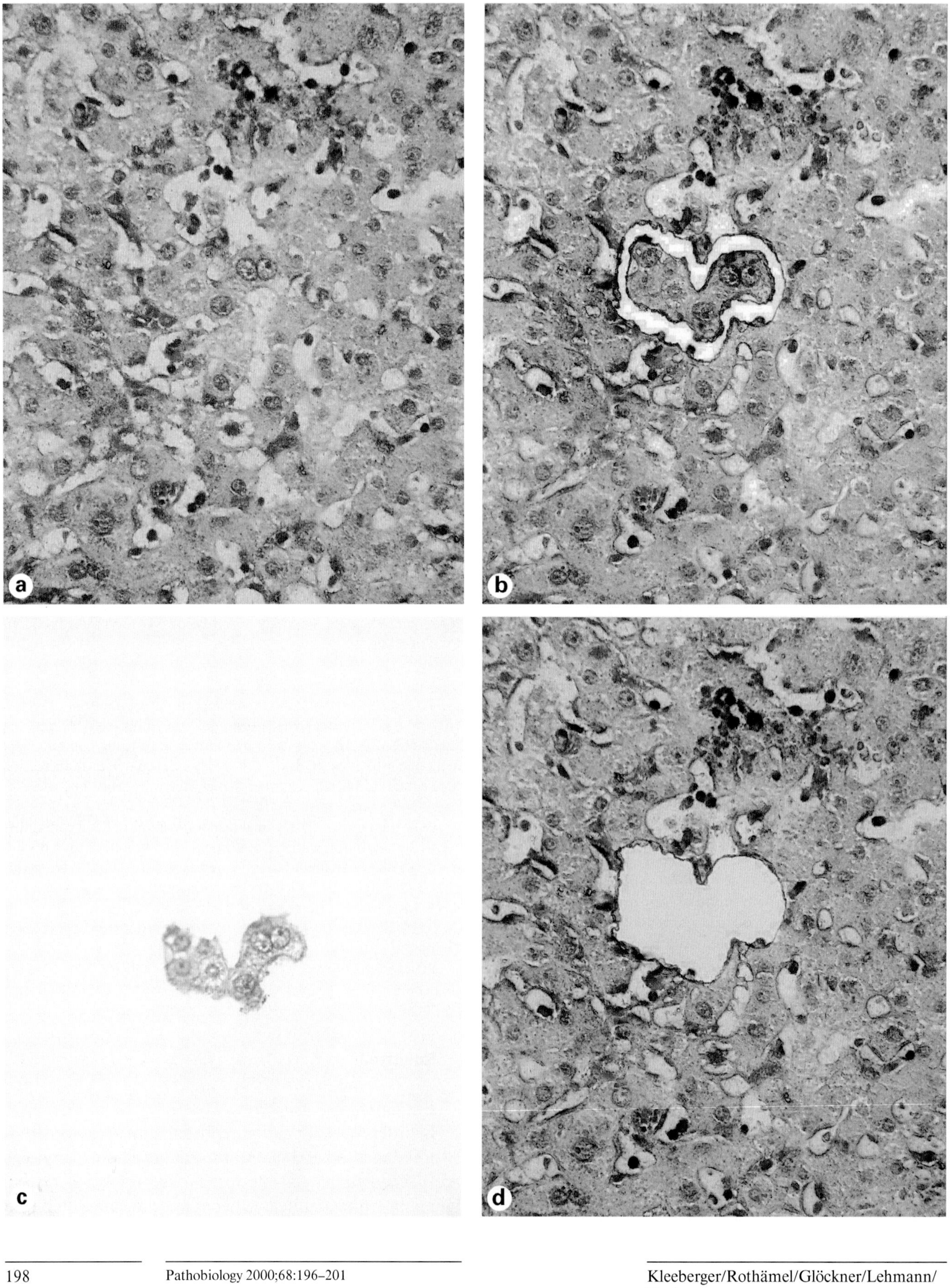

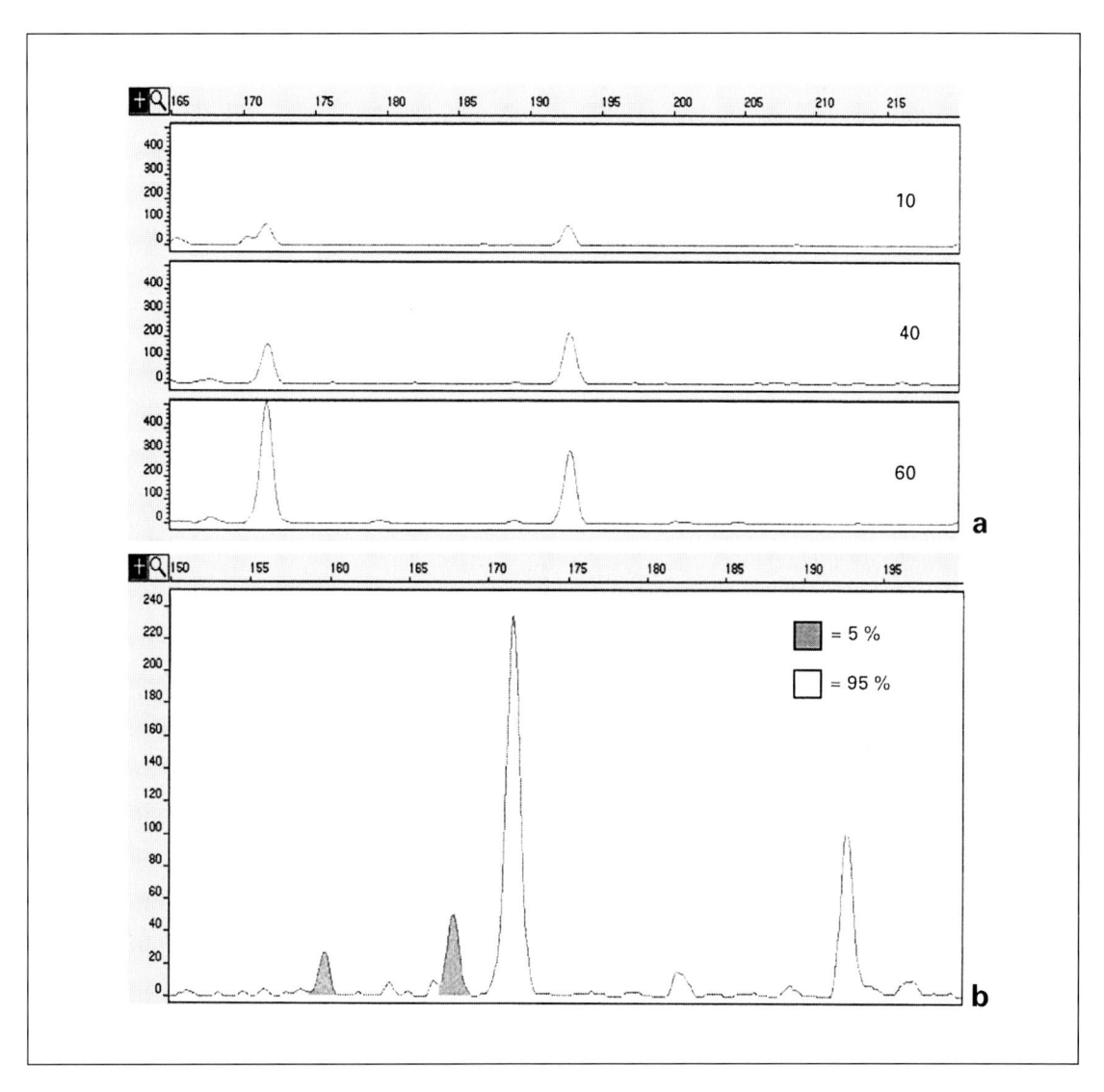

Fig. 3. a Defined cell counts from one tissue section ranging from 10 to 60 cells were microdissected. **b** Example of a mixing experiment with 5% cells of one individual mixed together with 95% cells of another individual in one sample. (Vertical scale represents peak height, horizontal scale fragment length.)

the lid. In several series of samples containing defined numbers of microdissected cells from the same tissue section ranging from 10 to 100 cells (fig. 3a), the genotype of the isolated cells could be reproducibly determined already in the samples containing 10 cells. Also in samples con-

Fig. 2. Laser-based microdissection of hepatic cells. Hematoxylin. ×630. **a** Tissue section of a posttransplantation liver biopsy. Lymphocytes and monocytes are immunohistochemically labeled (fast red). The laser beam cuts out the unlabeled target structures (**b**) which can then be catapulted into the lid of a PCR tube (**c**). **d** After that an empty space is left on the tissue section.

taining only 5 cells the alleles could be detected, but these results were not consistent and thus not reproducible. On the other hand, the peak height only roughly correlated with the number of cells pooled in the different samples, reflecting the fact that only cell cuts are isolated from the tissue sections which in part might not contain the DNA region analyzed in this PCR assay.

Mixing Experiments

Next we performed experiments where cells originating from different individuals were mixed together in certain ratios in order to find out to which degree a chimeric

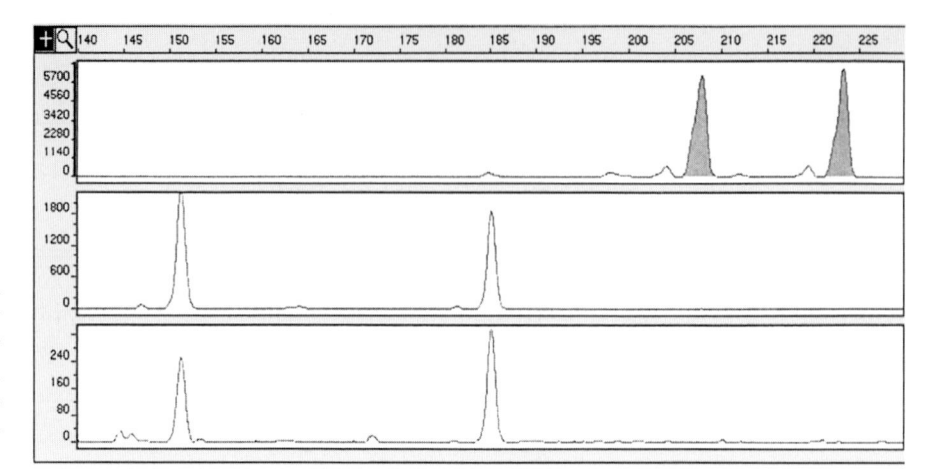

Fig. 4. PCR assay of endothelial cells from a posttransplantation heart biopsy. Upper row shows the recipient's genotype. Middle row shows the donor's genotype. Lower row with the PCR product of the microdissected endothelial cells exhibits only the donor's alleles.

state can be detected by this PCR assay. Figure 3b shows one example where 2 microdissected cells from one individual were mixed together with 38 cells from another individual, i.e. cells from different individuals were mixed in one sample in a ratio of 5 to 95%. Within such a ratio the 5% portion could clearly be determined indicating that a chimeric state with down to 5% cells of one individual within 95% of another can be detected by this PCR assay. Here again the peak heights of the different alleles did not relate to the number of sampled cells in a proportional manner.

Heart Biopsies

A total of 9 posttransplantation heart biopsies, all of them obtained within 4 weeks after transplantation and none of them showing histological signs of rejection, were chosen from the archive. Endothelial cells were immunostained with the CD31 antibody. As the endothelium is a morphologically very subtle structure which is sometimes difficult to recognize and as the resolution concerning morphology under the dissecting microscope is impaired these biopsies were used in order to test whether an isolation of endothelial cells without contamination by surrounding or infiltrating circulating blood cells, mainly lymphocytes, is technically possible. As expected no chimerism and thus no endothelial cells of a recipient origin could be detected, also demonstrating that no blood cells were catapulted into the samples by mistake (fig. 4).

Liver Biopsies

A total of 14 posttransplantation liver biopsies with the time interval between transplantation and biopsy taking ranging from 1 week to 1.5 years were retrieved from the archive. Histologically the livers showed various degrees of rejection and in few cases hepatitis or fibrosing cholangitis could be observed. Lymphocytes and monocytes were marked immunohistochemically using the CD45 together with the CD68 antibody. From all specimens periportal and centrilobular hepatocytes as well as bile duct cells were dissected and collected separately. In none of the samples could hepatocytes or bile duct cells of recipient origin be found.

Discussion

By combination of laser microdissection and STR-PCR for the investigation of intragraft microchimerism the cells of interest can be isolated under microscopic control. Target structures which are difficult to identify morphologically, such as endothelial cells, can be labeled immunohistochemically. Alternatively, possible surrounding and infiltrating circulating blood cells can be immunostained in order to avoid contamination with these cells by mistake. The subsequent highly sensitive PCR assay makes it possible to determine the genotype of donor and recipient in samples containing as few as 10 microdissected cells. Furthermore, a degree of chimerism of down to 5% of one genotype within 95% of another can be detected by this method.

Compared to the in situ hybridization methodologies that have been used to investigate intragraft chimerism in previous studies [6, 8, 9, 13] laser microdissection is able to isolate selectively the target structures after immunohistochemical labeling from tissue slides that preserve their morphological integrity under normal microscopic vision. The signals obtained after PCR are positively evaluated, i.e. only the existence and not the absence of a signal is considered for evaluation. In addition the PCR signals are directly evaluated, i.e. the results obtained need not to be divided by certain factors in order to level out false-negative and false-positive results.

The investigation of endothelial cells from heart biopsies which were obtained shortly after transplantation presenting no signs of rejection or other histopathological findings showed that in tissues where the endothelial cells are very likely to be of donor origin only, no cells of recipient origin were detected in the samples. Thus, the possibility of creating artificial chimerism due to contamination of the samples with blood cells lying in direct vicinity to the target structures can be avoided by exact microdissection.

In a total of 14 liver biopsies examined for the existence of chimerism on the epithelial level, no hepatocytes and bile duct cells of recipient origin could be detected. These findings do not contradict previous studies that were able to find recipient-derived hepatocytes using in situ hybridization of the X- and Y-chromosomes [8, 9], as our method is only able to detect a chimeric state with a sensitivity of down to 5% and as we cannot collect all liver epithelia contained in one tissue section. But it indicates that the integration of recipient-derived parenchymal cells into the transplanted liver at least does not play a major role in terms of quantity.

To the best or our knowledge this is the first study to use the combination of laser-assisted microdissection and STR-PCR of one highly polymorphic tetranucleotide marker for the investigation of a possible intragraft microchimerism on the epithelial and endothelial levels in grafts after solid organ transplantation. Although in a first series of liver and heart biopsies no hepatocytes and endothelial cells of recipient origin could be detected, further studies will be necessary to examine more cases and to correlate the results with parameters such as the time interval between transplantation and biopsy taking, degree of rejection or other pathological findings such as recurrence of the initial disease, particularly hepatitis C in livers. Another important issue will be to extend the investigations to other transplanted organs such as kidneys and lungs in order to compare the organ systems which may give further answers to the question of why these organs behave differently after transplantation.

References

1 Starzl TE, Demetris AJ, Trucco M, Murase N, Ricordi C, Ildstad S, Ramos H, Todo S, Tzakis A, Fung JJ, et al: Cell migration and chimerism after whole-organ transplantation: The basis of graft acceptance. Hepatology 1993;17:1127–1152.

2 Hisanaga M, Schlitt HJ, Hundrieser J, Nakajima Y, Kanehiro H, Nakano H, Pichlmayr R: Role of the graft as a source of donor-type microchimerism in liver transplant patients. Transplant Proc 1996;28:1073–1075.

3 Wood K, Sachs DH: Chimerism and transplantation tolerance: Cause and effect. Immunol Today 1996;17:584–588.

4 Rao AS, Starzl TE, Demetris AJ, Trucco M, Thomson A, Qian S, Murase N, Fung JJ: The two-way paradigm of transplantation immunology. Clin Immunol Immunopathol 1996;80: S46–S51.

5 Schlitt HJ: Is microchimerism needed for allograft tolerance? Transplant Proc 1997;29:82–84.

6 Andersen CB, Ladefoged SD, Larsen S: Cellular inflammatory infiltrates and renal cell turnover in kidney allografts: A study using in situ hybridization and combined in situ hybridization and immunohistochemistry with a Y-chromosome-specific DNA probe and monoclonal antibodies. APMIS 1991;99:645–652.

7 Petersen BE, Bowen WC, Patrene KD, Mars WM, Sullivan AK, Murase N, Boggs SS, Greenberger JS, Goff JP: Bone marrow as a potential source of hepatic oval cells. Science 1999;284: 1168–1170.

8 Theise ND, Nimmakayalu M, Gardner R, Illei PB, Morgan G, Teperman L, Henegariu O, Krause DS: Liver from bone marrow in humans. Hepatology 2000;32:11–16.

9 Alison MR, Poulsom R, Jeffery R, Dhillon AP, Quaglia A, Jacob J, Novelli M, Prentice G, Williamson J, Wright NA: Hepatocytes from nonhepatic adult stem cells. Nature 2000;406:257.

10 Asahara T, Masuda H, Takahashi T, Kalka C, Pastore C, Silver M, Kearne M, Magner M, Isner JM: Bone marrow origin of endothelial progenitor cells responsible for postnatal vasculogenesis in physiological and pathological neovascularization. Circ Res 1999;85:221–228.

11 Lin Y, Weisdorf DJ, Solovey A, Hebbel RP: Origins of circulating endothelial cells and endothelial outgrowth from blood. J Clin Invest 2000;105:71–77.

12 Ferrari G, Cusella-De Angelis G, Coletta M, Paolucci E, Stornaiuolo A, Cossu G, Mavilio F: Muscle regeneration by bone marrow-derived myogenic progenitors. Science 1998;279:1528–1530.

13 Bittmann I, Baretton GB, Schneeberger H: Chronic transplant reaction of the kidney. A interphase cytogenetic and immunohistologic characterization of the involved cells in relation to donor and recipient origin. Pathologe 1998;19:129–133.

14 Hruban RH, Long PP, Perlman EJ, Hutchins GM, Baumgartner WA, Baughman KL, Griffin CA: Fluorescence in situ hybridization for the Y-chromosome can be used to detect cells of recipient origin in allografted hearts following cardiac transplantation. Am J Pathol 1993; 142:975–980.

15 Lehmann U, Glöckner S, Kleeberger W, von Wasielewski HF, Kreipe H: Detection of gene amplification in archival breast cancer specimens by laser-assisted microdissection and quantitative real-time polymerase chain reaction. Am J Pathol 2000;156:1855–1864.

16 Scharf SJ, Smith AG, Hansen JA, McFarland C, Erlich HA: Quantitative determination of bone marrow transplant engraftment using fluorescent polymerase chain reaction primers for human identity markers. Blood 1995;85:1954–1963.

17 Thiede C, Florek M, Bornhauser M, Ritter M, Mohr B, Brendel C, Ehninger G, Neubauer A: Rapid quantification of mixed chimerism using multiplex amplification of short tandem repeat markers and fluorescence detection. Bone Marrow Transplant 1999;23:1055–1060.

18 Leclair B, Fregeau CJ, Aye MT, Fourney RM: DNA typing for bone marrow engraftment follow-up after allogeneic transplant: A comparative study of current technologies. Bone Marrow Transplant 1995;16:43–55.

19 Moller A, Brinkmann B: Locus ACTBP2 (SE33). Sequencing data reveal considerable polymorphism. Int J Legal Med 1994;106:262–267.

20 Polymeropoulos MH, Rath DS, Xiao H, Merril CR: Tetranucleotide repeat polymorphism at the human beta-actin related pseudogene H-beta-Ac-psi-2 (ACTBP2). Nucleic Acids Res 1992;20:1432.

Quantitative Molecular Analysis of Laser-Microdissected Paraffin-Embedded Human Tissues

Ulrich Lehmann   Oliver Bock   Sabine Glöckner   Hans Kreipe

Institute of Pathology, Medizinische Hochschule Hannover, Germany

Key Words

Biopsies, archival · Gene amplification · Laser microdissection · Methylation · PCR, real time · Reverse transcription PCR

Abstract

Laser microdissection enables the contamination-free isolation of morphologically defined pure cell populations from archival formalin-fixed paraffin-embedded tissue specimens. Cells isolated by this method have been characterized by a wide variety of qualitative molecular assays, e.g. loss of heterozygosity, point mutations, clonality and lineage origin. The recently introduced real-time PCR technology renders the reliable quantification of very small amounts of nucleic acids possible. Several groups including our own showed that this technique can be successfully applied for the quantification of DNA and RNA isolated from microdissected archival tissue sections, even after immunohistochemical staining. The exact analysis of quantitative changes of nucleic acids during the course of pathological alterations has thus become possible. In many situations these quantitative changes can be expected to be more important than qualitative changes. The new technology for the quantification of structural genomic alterations and changes in the gene expression pattern in conjunction with microdissection have equipped morphologists with a powerful tool to study reactive and neoplastic changes of tissues.

Copyright © 2001 S. Karger AG, Basel

A major methodological problem in the analysis of neoplastic tissues using the whole range of molecular biological techniques for the detection of qualitative and quantitative alterations in the genome of tumor cells or their mRNA population is the heterogeneity of the primary tumor samples. Molecular alterations like an increase in gene copy number or changes in gene expression activities can be masked by contaminating bystander cells. In order to overcome this obstacle various microdissection devices have been developed for the contamination-free isolation of homogeneous cell populations. Early systems were very time-consuming and required a great deal of skill and experience. The development of laser-assisted microdissection systems employing highly sophisticated technology for the transfer of isolated cells from the histological slide into a reaction tube has greatly facilitated progress in this area of research [1, 2]. Figure 1 illustrates the isolation of a single cell from a histological slide.

Microdissection Systems Available

Two different commercially available laser micordissection systems are used alternatively in most laboratories [1, 2]. Both have their advantages and disadvantages, depending on the application. The technical details of these two systems and others described in the literature are beyond the scope of this article. All systems in use

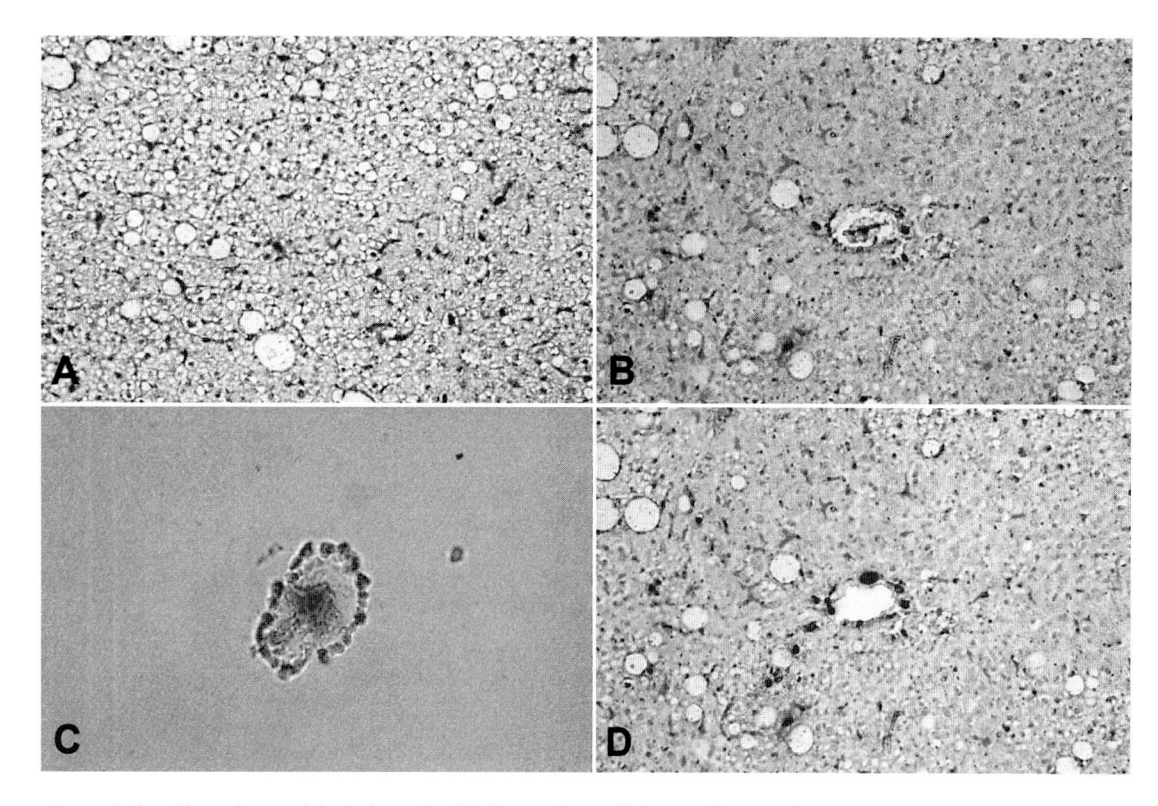

Fig. 1. Microdissection and isolation of a CD68-positive cell from a liver section. **A** Normal liver. Immunohistochemical detection of CD68. ×100. **B** CD68-positive cell is microdissected by the UV laser microbeam. ×100. **C** Isolated CD68-positive cell in the lid of a reaction tube. ×400. **D** Section after removal of CD68-positive cell. ×100. **A–D** HE counterstain.

seem to be compatible with a subsequent molecular analysis of isolated nucleic acids or proteins. A central problem of all systems known to the authors is the frequently occurring unsatisfactory conservation of morphological details during the process of microdissection. This problem represents currently one of the major drawbacks of this technology limiting its range of applications. Morphological identification of cells during microdissection can be improved by prior immunolabelling. Nevertheless, future improvements are definitely required.

Real-Time PCR Systems

Principle

The theory and practice of real-time PCR technology have been extensively discussed in several recent reviews [3, 4]. In this context it is only necessary to mention that in addition to the primers specific fluorescence-labelled probes are added to the reaction mixture. These probes are hydrolyzed during the extension phase of the reaction, thereby generating a fluorescence signal (hybridization probes, TaqMan technology) or generate a fluorescence signal only if hybridizing adjacent to each other to the PCR product (hybridization probes, LightCycler technology) (fig. 2). In both cases the fluorescence signal is directly proportional to the accumulation of the PCR product and measured online ('in real time') during the whole reaction process, thereby enabling an accurate and very fast determination of the starting copy number of the amplified sequence. In addition a careful analysis of the data gives valuable information about the reaction efficiencies.

Advantages

The advantages of the real-time PCR methodology are the broad linear range (five to seven orders of magnitude), which is much larger than for every conventional quantitative PCR protocol, the speed and high throughput of the system, the elimination of all postamplification steps, the exclusion of all plateau effects at the end of the PCR reaction from the quantitative analysis and the superior sen-

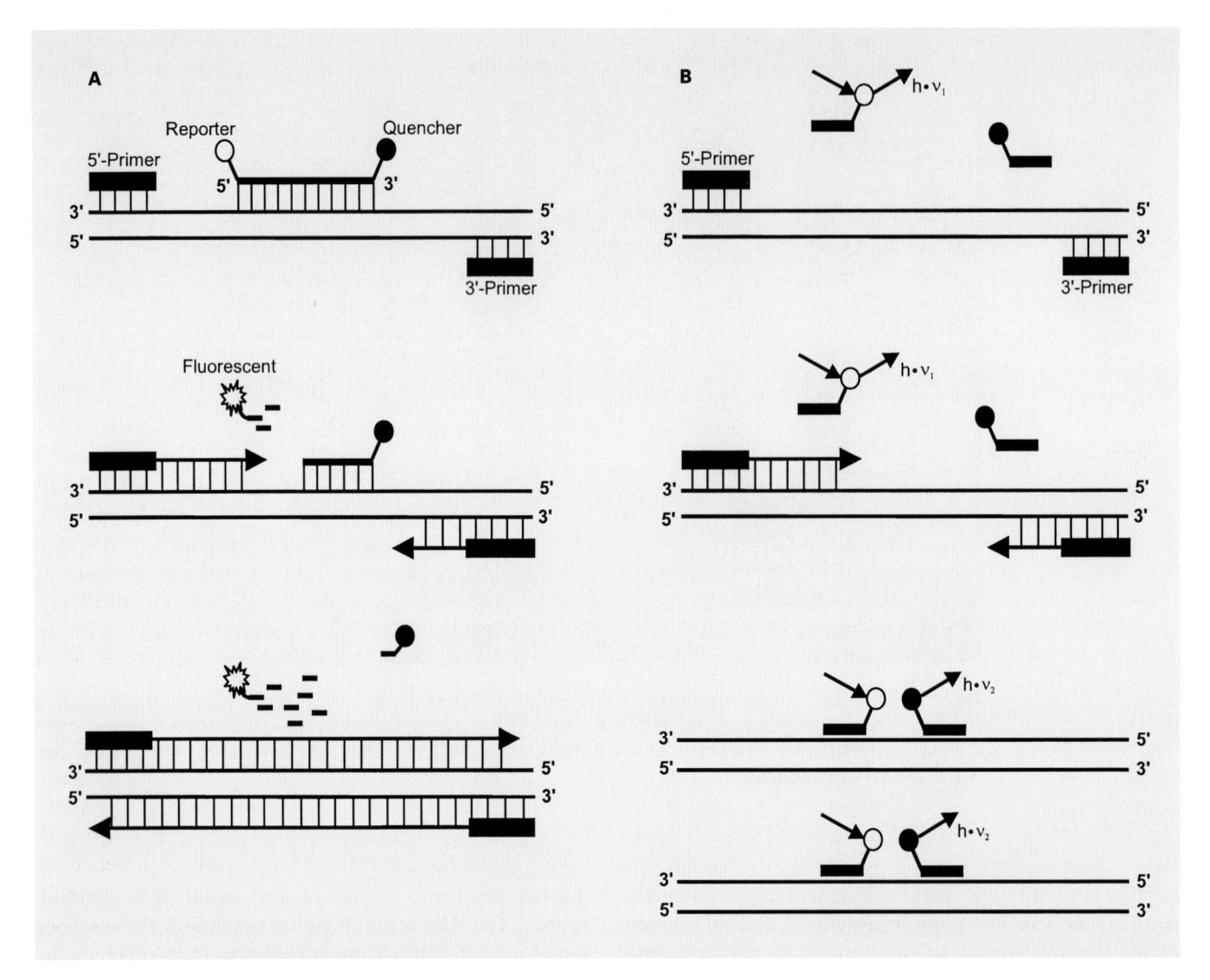

Fig. 2. Principle of **A** the hydrolysis assay (TaqMan™) and **B** the hybridization assay (LightCycler™) for the real-time PCR technology.

sitivity which comes close to the theoretical detection limit.

Conventional competitive and noncompetitive quantitative PCR assays require the analysis of a whole set of identical samples for a titration of the internal standard in order to quantitate the amount of target molecules in this sample. For this reason the analysis of microdissected samples is not possible, because they cannot be divided further into a set of five to six identical samples to perform the necessary titration of the internal standard. In addition, the linear range of these assays is much smaller than that of the kinetic real-time PCR, requiring extensive pretesting of samples to roughly determine the amount of target molecules, which is necessary for calibration of the internal standard. Again, this kind of pretesting is not possible analyzing the minute amounts of nucleic acids from formalin-fixed microdissected samples displaying a considerable heterogeneity with regard to DNA content.

Since the risk of contamination is a great problem working with minute amounts of starting material provided by laser-assisted microdissection the omission of all postamplification steps employing this technique is a great advantage.

For the accurate quantification of alterations in the gene copy number the real-time PCR technology – in contrast to the FISH technique – does not rely on long hybridization probes which are available only for a limited number of loci. There are also potentially important loci for which standardized manufactured hybridization probes suitable for the analysis of paraffin-embedded biopsies are not available (e.g. AIB-1 [5], CDK4 [6], PPABP [7]). Therefore, the flexibility and versatility of the primer and probe design offer a clear advantage especially in the area of research which should not be restricted by the availability of hybridization probes. In principle, for all loci in the human genome for which sequence information is available in databases a new quantitative real-time PCR assay can be developed and validated within 2 weeks.

Advantages of Relative Quantification

The real-time PCR technology is most often used for the absolute quantification of nucleic acids. However, the absolute numerical quantification of gene copy numbers has some disadvantages: (1) preparation of exact quantitative standards is laborious and a serious source of errors, (2) stable storage of quantitative standards is very difficult to achieve, (3) amplification of several standards for generating a calibration curve for absolute quantification within each run reduces the amount of samples which can be analyzed in each run, thereby diminishing the throughput of the system, and (4) the amplification of different template preparations (standard vs. sample) is compared. This last point is of particular importance concerning the amplification of DNA from pathological specimens derived from heterogeneous sources with different degrees of tissue preservation. Slightest impurities in the sample and partial template degradation due to improper fixation will severely distort the absolute quantification.

Application of Real-Time PCR Systems for the Analysis of Laser-Microdissected Samples

General Considerations

The most important parameter influencing the quantitative (but also the qualitative) molecular analysis is the process of specimen fixation. It critically determines the conservation and the integrity of the nucleic acids. Formalin fixation leads to a fragmentation of DNA and RNA. This makes the amplification of sequences longer than the average fragment size in the specimen very inefficient or in most cases impossible. In our experience the average fragment size of DNA in formalin-fixed specimens is between 200 and 600 bp, for RNA the fragments in most cases are smaller, between 150 and 300 bp.

The advantage of formalin fixation in addition to the better preserved morphological details is that the cross-linking of the nucleic acids protects them after deparaffinization against further degradation. This makes special staining procedures prior to the quantitative molecular analysis like immunohistochemistry more feasible and easier.

As already mentioned the fragmentation of DNA and RNA in the formalin-fixed specimens makes the amplification of longer fragments very inefficient. Using different primers generating amplicons of an increasing size from the same cDNA (see fig. 3) we could demonstrate that the actual reaction efficiency of the amplification process is not altered. But the amplification plots are shifted to the right (corresponding to weaker signals in a conventional PCR) because the concentration of templates which can generate the longer amplicons is reduced due to the fragmentation of the nucleic acids. That means that for longer amplicons the actual template concentration is reduced and the reaction efficiency remains unaltered. These results stress the importance of designing the primers and probes in such a way that the amplicon length is minimized.

Examples

The methodology described above can be used for the quantification of gene copy number alterations, the quantification of mRNA expression levels and the measurement of changes in the methylation pattern of promoter regions.

Gene Copy Number. A variety of human cancers carry specifically amplified oncogenes with a high copy number. The magnitude of amplification was found to correlate in different tumors with aggressive potential and proliferative activity. In breast cancer amplification and deletion are the most common mechanism leading to gene deregulation. In addition to well-established oncogenes like c-*myc*, *cyclinD1* or c-*erb-B2* a wide variety of loci are amplified. Yet the significance of most of these findings has to be established employing standardized detection methodologies. For this reasons we have combined laser-assisted microdissection of tumor cells and real-time PCR technology in order to quantitate objectively gene copy number alterations in pure morphologically defined tumor cell populations [8]. Following the arguments mentioned above concerning absolute and relative quantification strategies we used a relative quantification algorithm comparing normal and neoplastic tissue. As a reference we try to use normal tissue from the very same section or from a different section of the same patient. If this is not

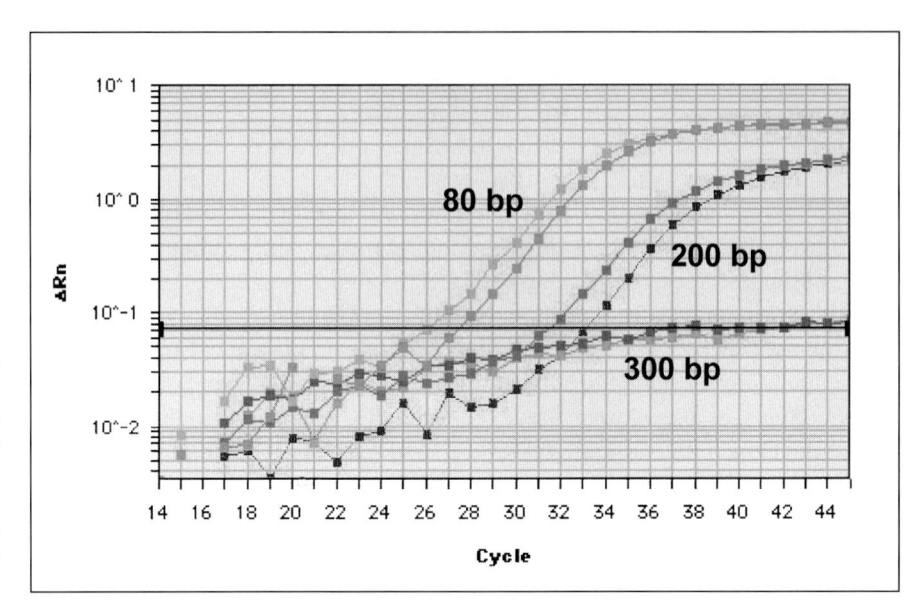

Fig. 3. Effect of amplicon size on amplification efficiency. The amplification plots for different real-time PCR assays using the same cDNA preparation as a template are shown. The sense primer and hybridization probe is also the same in all three reactions; only the antisense primer is different, thereby producing the indicated different amplicon sizes.

available we use as a standard the data obtained from the analysis of a series of 60 mammary epithelial samples microdissected from archival biopsies. This reference series represents the range of variability encountered due to different fixation and tissue processing conditions.

We are currently using this technique for a direct correlation of a morphological classification system of intraductal carcinomas of the mammary gland and the exact quantification of gene copy numbers in these lesions [9].

Expression Levels. Real-time PCR technology is most widely used for the quantitative analysis of gene expression. In the vast majority of published studies cell culture systems are used as a source of mRNA, a situation in which the amount and conservation of nucleic acids is not a problem at all for the quantitative analysis. Recently several groups reported the quantification of mRNA expression levels from fresh frozen biopsies. In some cases the authors also microdissected groups of cells from histologically stained frozen sections or even pooled single-picked cells [10]. In one study the mRNA isolated from microdissected cells was successfully hybridized to a cDNA array to monitor the expression of many genes in parallel [11]. For this purpose it was necessary to preamplify the cDNA after the first reverse transcription. Future studies have to demonstrate the reliability and general applicability of this approach. The extraction and qualitative detection of RNA from archival material have been reported independently in the last years by several groups [12–16]. But only very recently has the exact quantitative analysis of gene expression in formalin-fixed paraffin-embedded biopsies been reported. So far only whole sections have been analyzed [17, 18]. Since the analysis of gene expression in small and preneoplastic lesions is of uppermost importance for a better understanding of the biology of malignant transformation and because these lesions are almost exclusively detected in formalin-fixed paraffin-embedded specimens reliable protocols for the quantitative analysis of mRNA levels in microdissected cells have to be developed.

Methylation Analysis. A growing body of evidence accumulating in the last years has convincingly demonstrated that the methylation of cytosin residues in the promoter regions of several important genes is a frequently occurring event during the process of malignant transformation leading to the inactivation of growth and invasion-suppressing genes [19, 20 and references therein].

However, considering the discovery of methylation events also in normal tissue [see for example 21, 22] it is absolutely necessary to quantitate the extent of CpG methylation in the neoplastic tissue analyzed in order to get meaningful information about the alteration of methylation patterns during the clonal evolution of a neoplastic lesion. Unfortunately, the widely used term 'hypermethylation' is in most cases only very loosely defined. Different methods have been developed to quantitate the extent of methylation in a given sample. But surprisingly only very few reports about the application of these techniques exist in the literature. Also only very few publications compare different approaches to the detection and quantification of this important epigenetic alteration [23].

The very interesting and promising Ms-SnuPE assay has been used so far only by the group which developed the technique [24–26]. Three groups have independently developed a new methylation-specific PCR assay based on the TaqMan technology [27–29] (fig. 4). The advantage of this assay is the opportunity to directly quantitate the extent of methylation even in very small amounts of DNA. Therefore, this technique can be combined with laser-assisted microdissection. These tools can now be used for the systematic study of alterations of DNA methylation during the process of malignant progression and these data can be correlated with the morphological characterization of the lesions analyzed.

Future Directions

The development of protocols for the amplification of RNA isolated from microdissected formalin-fixed paraffin-embedded cells prior to the quantitative analysis has already started but awaits further improvements and confirmation of reliability [30]. Rigorous tests have to ensure that the amplification step does not distort the relative abundance of the original mRNA species.

If this technology is widely available a large scale expression analysis of small and precancerous lesions from archival biopsies can be initiated. The next step in the near future will be the application of the proteomics technology to microdissected biopsies. Probably the process of fixation will be a much greater hindrance for the analysis of proteins than for nucleic acid analysis.

But since several technologies available today could not even be envisioned a few years ago we should not be too pessimistic concerning the future developments.

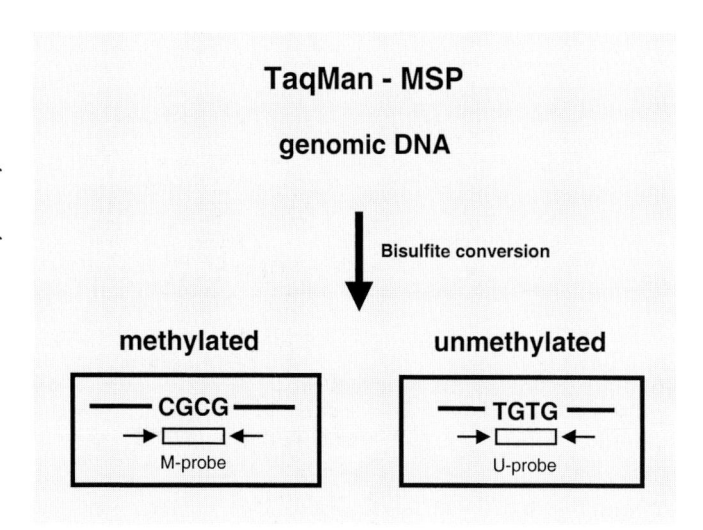

Fig. 4. Principle of the TaqMan-MSP assay: the detection of methylated and unmethylated DNA after bisulfite conversion in a quantitative manner using hybridization probes. The same methylation-insensitive primer pair is used in two different reactions containing a probe specific for methylated DNA (M-probe) or a probe specific for unmethylated DNA (U-probe).

References

1 Emmert-Buck MR, Bonner RF, Smith PD, Chuaqui RF, Zhuang Z, Goldstein SR, Weiss RA, Liotta LA: Laser capture microdissection. Science 1996;274:998–1001.
2 Schutze K, Lahr G: Identification of expressed genes by laser-mediated manipulation of single cells. Nat Biotechnol 1998;16:737–742.
3 Wittwer CT, Ririe KM, Andrew RV, David DA, Gundry RA, Balis UJ: The LightCycler: A microvolume multisample fluorimeter with rapid temperature control. Biotechniques 1997;22:176–181.
4 Lie YS, Petropoulos CJ: Advances in quantitative PCR technology: 5′ nuclease assays. Curr Opin Biotechnol 1998;9:43–48.
5 Anzick SL, Kononen J, Walker RL, Azorsa DO, Tanner MM, Guan XY, Sauter G, Kallioniemi OP, Trent JM, Meltzer PS: AIB1, a steroid receptor coactivator amplified in breast and ovarian cancer. Science 1997;277:965–968.
6 An HX, Beckmann MW, Reifenberger G, Bender HG, Niederacher D: Gene amplification and overexpression of CDK4 in sporadic breast carcinomas is associated with high tumor cell proliferation. Am J Pathol 1999;154:113–118.
7 Zhu Y, Qi C, Jain S, Le Beau MM, Espinosa R 3rd, Atkins GB, Lazar MA, Yeldandi AV, Rao MS, Reddy JK: Amplification and overexpression of peroxisome proliferator-activated receptor binding protein (PBP/PPARBP) gene in breast cancer. Proc Natl Acad Sci USA 1999;96:10848–10853.
8 Lehmann U, Glöckner S, Kleeberger W, von Wasielewski HF, Kreipe H: Detection of gene amplification in archival breast cancer specimens by laser-assisted microdissection and quantitative real-time polymerase chain reaction. Am J Pathol 2000;156:1855–1864.
9 Glöckner S, Lehmann U, Wilke N, Kleeberger W, Länger F, Kreipe H: Amplification of growth regulatory genes in intraductal breast cancer is associated with higher nuclear grade but not with the progression to invasiveness. Lab Invest, in press.
10 Fink L, Seeger W, Ermert L, Hanze J, Stahl U, Grimminger F, Kummer W, Bohle RM: Real-time quantitative RT-PCR after laser-assisted cell picking. Nat Med 1998;4:1329–1333.
11 Luo L, Salunga RC, Guo H, Bittner A, Joy KC, Galindo JE, Xiao H, Rogers KE, Wan JS, Jackson MR, Erlander MG: Gene expression profiles of laser-captured adjacent neuronal subtypes. Nat Med 1999;5:117–122.
12 Jackson DP, Lewis FA, Taylor GR, Boylston AW, Quirke P: Tissue extraction of DNA and RNA and analysis by the polymerase chain reaction. J Clin Pathol 1990;43:499–504.

13 Stanta G, Schneider C: RNA extracted from paraffin-embedded human tissues is amenable to analysis by PCR amplification. Biotechniques 1991;11:304, 306, 308.
14 Finke J, Fritzen R, Ternes P, Lange W, Dolken G: An improved strategy and a useful housekeeping gene for RNA analysis from formalin-fixed, paraffin-embedded tissues by PCR. Biotechniques 1993;14:448–453.
15 Ruster B, Zeuzem S, Roth WK: Hepatitis C virus sequences encoding truncated core proteins detected in a hepatocellular carcinoma. Biochem Biophys Res Commun 1996;219: 911–915.
16 Mizuno T, Nagamura H, Iwamoto KS, Ito T, Fukuhara T, Tokunaga M, Tokuoka S, Mabuchi K, Seyma T: RNA from decades-old archival tissue blocks for retrospective studies. Diagn Mol Pathol 1998;7:202–208.
17 Sheils OM, Sweeney EC: TSH receptor status of thyroid neoplasms – TaqMan RT-PCR analysis of archival material. J Pathol 1999;188: 87–92.
18 Godfrey TI, Kim S-H, Chavira M, Ruff DW, Warren RS, Gray JW, Jensen RH: Quantitative mRNA expression analysis from formalin-fixed, paraffin-embedded tissues using 5′ nuclease quantitative reverse transcription-polymerase chain reaction. J Mol Diagn 2000;2:84–91.
19 Jones PA, Laird PW: Cancer epigenetics comes of age. Nat Genet 1999;21:163–167.
20 Momparler RL, Bovenzi V: DNA methylation and cancer. J Cell Physiol 2000;183:145–154.
21 Salem C, Liang G, Tsai YC, Coulter J, Knowles MA, Feng AC, Groshen S, Nichols PW, Jones PA: Progressive increases in de novo methylation of CpG islands in bladder cancer. Cancer Res 2000;60:2473–2476.
22 Ueki T, Toyota M, Sohn T, Yeo CJ, Issa JP, Hurban RH, Goggins M: Hypermethylation of multiple genes in pancreatic adenocarcinoma. Cancer Res 2000;60:1835–1839.
23 Gonzalgo ML, Bender CM, You EH, Glendening JM, Flores JF, Walker GJ, Hayward NK, Jones PA, Fountain JW: Low frequency of p16/CDKN2A methylation in sporadic melanoma: Comparative approaches for methylation analysis of primary tumors. Cancer Res 1997;57:5336–5347.
24 Gonzalgo ML, Jones PA: Rapid quantitation of methylation differences at specific sites using methylation-sensitive single nucleotide primer extension (Ms-SNuPE). Nucleic Acids Res 1997;25:2529–2531.
25 Nguyen TT, Nguyen CT, Gonzales FA, Nichols PW, Yu MC, Jones PA: Analysis of cyclin-dependent kinase inhibitor expression and methylation patterns in human prostate cancers. Prostate 2000;43:233–242.
26 Nguyen TT, Mohrbacher AF, Tsai YC, Groffen J, Heisterkamp N, Nichols PW, Yu MC, Lubbert M, Jones PA: Quantitative measure of c-abl and p15 methylation in chronic myelogenous leukemia: Biological implications. Blood 2000;95:2990–2992.
27 Lo YM, Wong IH, Zhang J, Tein MS, Ng MH, Hjelm NM: Quantitative analysis of aberrant p16 methylation using real-time quantitative methylation-specific polymerase chain reaction. Cancer Res 1999;59:3899–3903.
28 Eads CA, Danenberg KD, Kawakami K, Saltz LB, Blake C, Shibata D, Danenberg PV, Laird PW: MethyLight: A high-throughput assay to measure DNA methylation. Nucleic Acids Res 2000;28:E32.
29 Lehmann U, Hasemeier B, Lilischkis R, Kreipe H: Quantitative analysis of promotor hypermethylation in laser-microdissected archival specimens. Lab Invest, in press.
30 Lockhart DJ, Winzeler EA: Genomics, gene expression and DNA arrays. Nature 2000;405: 827–836.

Laser Capture Microdissection: Methodical Aspects and Applications with Emphasis on Immuno-Laser Capture Microdissection

Falko Fend[a]   Marcus Kremer[a,b]   Leticia Quintanilla-Martinez[b]

[a]Institute of Pathology, Technical University Munich, and [b]GSF Research Center Neuherberg, Oberschleissheim, Germany

Key Words

Laser capture microdissection · Immunohistochemistry · RT-PCR · Gene expression · Polymerase chain reaction

Abstract

Laser capture microdissection (LCM) is an easy, extremely fast and versatile method for the isolation of morphologically defined cell populations from complex primary tissues for molecular analyses. However, the optical resolution is limited due to the use of dried sections without coverslip necessary for tissue capture, and routine stains such as hematoxylin and eosin are sometimes insufficient for precise microdissection, especially in tissues with diffuse intermingling of different cell types and lack of easily discernible architectural features. Therefore, several groups have adapted immunohistochemical staining techniques for LCM. In addition to providing high contrast targets for microdissection, immunostaining allows selection of cells not only according to morphological, but also phenotypical and functional criteria. In order to allow reliable tissue transfer on one hand and preserve the integrity of the target of analysis such as DNA, RNA and proteins on the other hand, immunostaining protocols have to be modified for the purposes of LCM. The following review gives an overview of immuno-LCM and describes some applications, e.g. in the field of hematopathology.

Copyright © 2001 S. Karger AG, Basel

Introduction

The molecular analysis of pathologically altered primary tissues has brought significant advances for our understanding of disease mechanisms and also resulted in the development of a whole array of new diagnostic tests with a significant impact on patient management and therapy. This is especially true of human neoplasms, and the advent of high-throughput analytical tools such as cDNA arrays will allow to establish individual molecular profiles of tumors, complementing and extending the diagnostic and prognostic information gained from conventional histopathological examination. However, surgically obtained specimens of tumors are a complex mixture of neoplastic cells and reactive cellular elements, and the reactive component frequently outnumbers the tumor cell population. Whereas some molecular tests, such as the detection of disease-specific chromosomal translocations by Southern blot or PCR, are moderately to highly sensitive and can detect a small minority of cells carrying the alteration in question, others such as the detection of loss of heterozygosity or identification of point mutations in oncogenes or tumor suppressor genes by direct sequencing of PCR products will yield false-negative results if contamination by reactive cells reaches a certain threshold. On the RNA and protein level, an assignment of expressed genes and proteins to specific cell populations may be impossible if bulk tissue is used.

In the light of these problems, microdissection techniques were introduced as tools for obtaining homogeneous cell populations from complex primary tissues [1–7]. Recently, the development of laser-based dissection technologies such as laser capture microdissection (LCM) or laser microbeam microdissection with laser pressure catapulting has resulted in a breakthrough in terms of speed, versatility and ease of use [8–13]. This transformed high-precision microdissection from a time-consuming technique restricted to highly skilled workers in devoted research labs into a simple, standard procedure easily applicable in any pathology laboratory.

However, one of the drawbacks of most laser-assisted dissection devices including LCM is the necessity for the use of dehydrated sections without coverslip, which leads to a significant decrease in optical resolution. Although the morphological details of routinely stained sections (e.g. hematoxylin-eosin, HE) are sufficient for many purposes, precise isolation of homogeneous cell populations from complex tissues lacking easily discernible architectural features such as lymphoma, inflammatory infiltrates or diffusely infiltrating carcinomas can be virtually impossible. Immunohistochemical and cytochemical stains potentially are an important path to circumvent this problem, since they can render easily discernible, high-contrast targets. Furthermore, they could allow the isolation of cell populations according to phenotypic and functional criteria, complementing and expanding morphology. However, the requirements of LCM on one hand and the desire for optimal preservation of the target of analysis (DNA, RNA or protein) on the other sometimes require significant adaptations of conventional immunohistochemical staining techniques.

DNA Analysis

DNA is currently still the most frequent substrate for molecular examinations in pathology. Although high molecular weight DNA can only be obtained from fresh or frozen tissue and nucleic acids are fragmented to a significant degree in conventionally formalin-fixed, paraffin-embedded tissue, these archival sources can still be used for a wide range of PCR-based tests. In our experience and that of others, conventional immunostaining protocols for paraffin-embedded tissues do not significantly influence DNA quality or tissue capture by LCM and can be used without major modifications [14]. Already before the introduction of laser-assisted microdissection, immunostains or nonradioactive in situ hybridization have been used successfully to identify and recover target cells from tissue sections for subsequent molecular analysis [15]. An example is the isolation of the neoplastic Reed-Sternberg cells of Hodgkin's disease by micromanipulation of immunostained single cells. The amplification of identical, clonal immunoglobulin gene rearrangements from multiple single cells dissected from frozen sections finally confirmed their derivation from germinal center B cells with the exception of rare cases of T cell origin [16–18].

The application of no-touch laser-assisted microdissection techniques for single-cell procurement should result in a significant reduction of dissection time without compromising dissection precision. Although it was not primarily designed for single cell capture, LCM can be used to pool larger numbers of single cells on one dissection cap for subsequent analysis, in contrast to the cell-by-cell technique used with micromanipulation [19–21]. We have recently used this approach for the analysis of the clonal relationship between the neoplastic cells of Hodgkin's disease and a cutaneous T cell lymphoma arising in the same patient. Cloning and sequencing of PCR products obtained from different groups of RS cells identified by CD30 immunostaining revealed identical Ig heavy chain gene rearrangements in all PCR reactions, confirming them to be clonal B cells unrelated to the neoplastic T cell clone [22]. We used the same strategy to analyze the clonality of EBV-infected, B cell marker-positive RS cells arising in the background of a peripheral T cell lymphoma. The presence of multiple bands of different sizes obtained from groups of immunostained RS cells indicated that this phenomenon represents an expansion of EBV-infected reactive B cells rather than a true composite lymphoma [21].

The analysis of groups of isolated cells can save time and cost by significantly reducing the amount of necessary PCR reactions. Although it carries a higher risk of contamination by unwanted cells, we are confident that the pooling of single cells by LCM is of sufficient precision for most analyses. In addition, it may have further advantages by reducing artifacts due to sectioned nuclei or excessive numbers of amplification cycles.

Furthermore, recent technical developments for LCM such as cap surfaces not in contact with the tissue section, special extraction chambers for small fluid volumes or the membrane-covered, rotating cone replacing the conventional LCM cap developed by Suarez-Quian et al. [23] will further improve the precision of single cell microdissection.

Apart from Hodgkin's disease and related disorders, both normal and neoplastic lymphoid tissues in general are good examples for the necessity of phenotype-based microdissection, since they usually contain an intricate mixture of different subsets of lymphocytes lacking discriminating morphological features. We used LCM of immunostained paraffin sections to examine clonality in rare cases of composite B cell lymphomas showing two morphologically and phenotypically distinct cell populations [14]. PCR amplification of rearranged immunoglobulin genes was performed on DNA obtained both from gross tissue as well as on the two distinct cell populations microdissected from immunostained tissue sections. Whereas gross tissue-derived DNA showed a single band in all cases, PCR of the microdissected specimens amplified two distinct bands derived from two unrelated B cell clones as confirmed by sequencing. Since both populations were present in the gross tissue in roughly equivalent amounts, one would expect to be able to amplify both rearrangements simultaneously with the consensus primers used in this study. However, preferential amplification of one clone occurred in gross tissue extracts, possibly due to different priming efficiencies, wrongly suggesting monoclonality. These experiences indicate that microdissection to obtain purer cell populations may also be beneficial for analyses in which the percentage of the target population may seem high enough for conventional examination of bulk tissue.

In addition to the analysis of single or few genes by PCR, DNA obtained from microdissected paraffin-embedded tissue sections can be used for genome-wide screening techniques such as comparative genomic hybridization or genome-wide loss of heterozygosity screening, after a step of whole genome amplification if necessary [24, 25]. Further studies are needed to confirm the representativity of the amplified material in comparison to native DNA when a preanalysis random amplification step is used, and to establish the optimal conditions and the minimal amount of cells needed for reproducible results.

Analysis of Gene Expression

The establishment of gene expression signatures for different normal and pathologically altered tissues will enhance our abilities to understand and classify human disease and might provide us with tools for better prognostication and more refined, individually tailored treatments. However, assignment of expressed genes to the various cell populations present in heterogeneous primary tissues can be difficult or impossible, and quantification of RNA expression frequently is strongly influenced by the variable prevalence of the target population (e.g. tumor cells). Confirmation by in situ techniques such as immunohistochemistry or in situ hybridization may not always be possible, especially for low abundance transcripts, and is cumbersome and time-consuming when a large number of transcripts have to be examined. Therefore, many groups have tried to develop microdissection protocols that yield RNA of sufficient quality for downstream applications such as RT-PCR, expression library construction and cDNA array hybridization [9, 11, 13, 19, 26–37]. However, mRNA extraction poses more stringent requirements to tissue preservation and handling due to its higher sensitivity to fixation and degradation by ubiquitously present RNases unless stringent RNase-free conditions are observed. Nevertheless, several groups have demonstrated that microdissected frozen tissue can render good quality RNA from microdissected frozen tissue. Laser-based techniques have the advantage of performing the microdissection on completely dehydrated tissue sections or cell preparations, thus probably blocking endogenous RNases, and being significantly faster than manual or micromanipulation-based approaches. Cells isolated by LCM or laser microbeam microdissection are suitable mRNA sources for quantitative real-time PCR, both fluorescent or radioactively labelled probes for cDNA array hybridization, or expression library construction [19, 27, 30, 31, 34, 35, 38]. Linear amplification with T7 RNA polymerase has been shown to generate sufficient template for cDNA array hybridization [19]. As many as 5,000 cells are sufficient for generating reproducible results with radioactively labelled probes hybridized to commercial nylon filter arrays [32]. Although fresh frozen tissue remains the source of choice for RNA extraction, paraffin-embedded, formalin-fixed tissue can be used for certain applications such as RT-PCR, if the size of the chosen amplicons is small. Using a sensitive nested RT-PCR approach, Schütze and Lahr [11] have amplified expressed gene fragments from single cells microdissected from paraffin tissue sections. However, using such small amounts of starting material, the potential for technical artifacts caused by sectioning, contamination through fragments of attached cells and the high numbers of amplification cycles has to be kept in mind. Specht et al. [39] recently demonstrated the feasibility of quantitative real-time RT-PCR from microdissected formalin-fixed, paraffin-embedded tissue sections. Highly reproducible results were obtained with small amplicon sizes down to a level of approximately 2,000 cells, with an increasing variation below that number.

For visualization of target cell populations sufficient for many types of tissues, conventional stains such as HE are fast to perform, and do not lead to a major loss or

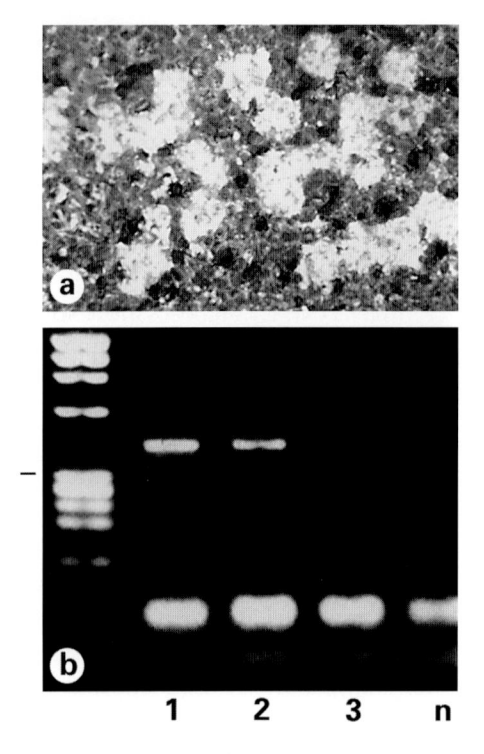

Fig. 1. a LCM of a frozen section of a reactive lymph node immunostained for CD3. The holes created by the LCM procedure are clearly visible. The immunostained T cells are left behind. **b** RT-PCR amplification of a 424-bp fragment of CD19 mRNA from a larger area of microdissected B cells after immuno-LCM. Lanes 1–3 show products with undiluted cDNA, a 1:5 dilution and a 1:25 dilution. N = Negative control.

RNA. If for reasons outlined above a higher level of optical resolution or identification of phenotypically diverse but morphologically similar cell types are desirable, immunolabelling can be applied.

In contrast to immunohistochemistry on paraffin sections, application of immunological staining techniques to frozen sections or cell preparations suitable for subsequent LCM requires a significant deviation from standard staining protocols and always results in a reduction of the RNA recovery compared to conventional stains. Reduction of staining times to less than 15 min in aqueous media and RNase-free conditions preserves sufficient high-quality RNA, allowing the amplification of cell-specific mRNA of more than 400 bp from captured cells with conventional single-step RT-PCR (fig. 1) [40]. Jin et al. [20] amplified specific mRNAs encoding for pituitary hormones from single immunostained cells isolated by LCM from cytospins with a sensitive nested RT-PCR. In addition to RNA recovery, a second crucial point for successful LCM from immunostained frozen sections is transfer efficiency, since frozen sections tend to adhere strongly to the glass slides, especially after drying steps. Good tissue transfer usually can be ensured by avoiding prolonged drying of sections, careful complete dehydration or pretreatment with glycerol [20, 40].

In order to further reduce the time of exposure to aqueous media and thus improve RNA recovery, Murakami et al. [41] have developed an ultrafast immunofluorescence staining technique which relies on the detection of weak fluorescent signals not visible by conventional means with the help of a very sensitive CCD camera. This makes possible staining times of 1 min, bringing them into the time span of conventional staining procedures.

If the problem of RNA recovery from immunostained sections can be resolved satisfactorily, gene expression can be correlated with phenotypic and functional properties of the examined cell population, such as proliferation, maturation stage and oncoprotein expression. Approaches such as RT-PCR or quantitative RT-PCR are probably more suited for partially degraded samples such as cells obtained by immuno-LCM rather than screening techniques like cDNA array hybridization, which are more likely to be influenced by partial degradation and bias introduced through template amplification.

Outlook

The spread of laser-assisted microdissection techniques will speed up identification of molecular alterations associated with human disease. The ability to separately analyze heterogeneous populations in complex primary tissues allows us to retrace the genetic evolution of neoplasms. The accumulation of genetic changes from metaplasia, to preinvasive changes and finally to invasive carcinoma can elegantly be demonstrated on microdissected tissue areas from a single case [7, 42].

The application of proteomics to microdissected tissues has opened a new bridge to 'molecular morphology'. The feasibility of two-dimensional gel electrophoresis, immunoblotting and immunoassays performed on cells obtained by LCM has been demonstrated by several groups [43–48]. The development of protocols to optimize recovery of nucleic acids and proteins from various microdissected sources, fast immunostaining methods for phenotype-directed microdissection and expression analysis and technical developments in terms of increased precision and automatization of laser-assisted microdissection will have a significant impact on tissue-based research and diagnostics.

References

1 Bianchi AB, Navone NM, Conti CJ: Detection of loss of heterozygosity in formalin-fixed, paraffin-embedded tumor specimens by the polymerase chain reaction. Am J Pathol 1991;138: 279–284.
2 Deng G, Lu Y, Zlotnikov G, Thor AD, Smith HS: Loss of heterozygosity in normal tissue adjacent to breast carcinomas. Science 1996; 274:2057–2059.
3 Küppers R, Zhao M, Hansmann ML, Rajewsky K: Tracing B cell development in human germinal centres by molecular analysis of single cells picked from histological sections. EMBO J 1993;12:4955–4967.
4 Whetsell L, Maw G, Nadon N, Ringer PD, Schaefer FV: Polymerase chain reaction microanalysis of tumors from stained histological slides. Oncogene 1992;7:2355–2361.
5 Going JJ, Lamb RF: Practical histological microdissection for PCR analysis. J Pathol 1996; 179:121–124.
6 Zhuang Z, Bertheau P, Emmert-Buck MR, Liotta LA, Gnarra J, Lineham WM, Lubensky IA: A microdissection technique for archival DNA analysis of specific cell populations in lesions <1 mm in size. Am J Pathol 1995;146: 620–625.
7 Walch A, Komminoth P, Hutzler P, Aubele M, Höfler H, Werner M: Microdissection of tissue sections: Application to the molecular genetic characterization of premalignant lesions. Pathobiology 2000;68:9–17.
8 Becker I, Becker K-F, Röhrl MH, Minkus G, Schütze K, Höfler H: Single-cell mutation analysis of tumors from stained histologic slides. Lab Invest 1997;75:801–807.
9 Bernsen MR, Dijkman HBPM, de Vries E, Figdor CG, Ruiter DJ, Adema GJ, van Muijen GNP: Identification of multiple mRNA and DNA sequences from small tissue samples isolated by laser-assisted microdissection. Lab Invest 1998;78:1267–1273.
10 Böhm M, Wieland I, Schütze K, Rübben H: Microbeam MOMeNT: Non-contact laser microdissection of membrane-mounted native tissue. Am J Pathol 1997;151:63–67.
11 Schütze K, Lahr G: Identification of expressed genes by laser-mediated manipulation of single cells. Nat Biotechnol 1998;16:737–742.
12 Bonner RF, Emmert-Buck M, Cole K, Pohida T, Chuaqui R, Goldstein S, Liotta LA: Laser capture microdissection: Molecular analysis of tissue. Science 1997;278:1481–1483.
13 Emmert-Buck MR, Bonner RF, Smith PD, Chuaqui R, Zhuang Z, Goldstein SR, Weiss RA, Liotta LA: Laser capture microdissection. Science 1996;274:998–1001.
14 Fend F, Quintanilla-Martinez L, Kumar S, Beaty MW, Blum L, Sorbara L, Jaffe ES, Raffeld M: Composite low grade B-cell lymphomas with two immunophenotypically distinct cell populations are true biclonal lymphomas. A molecular analysis using laser capture microdissection. Am J Pathol 1999;154:1857–1866.

15 d'Amore F, Stribley JA, Ohno T, et al: Molecular studies on single cells harvested by micromanipulation from archival tissue sections previously stained by immunohistochemistry or nonisotopic in situ hybridization. Lab Invest 1997;76:219–224.
16 Küppers R, Rajewsky K, Zhao M, et al: Hodgkin's disease: Hodgkin and Reed-Sternberg cells picked from histological sections show clonal immunoglobulin gene rearrangements and appear to be derived from B cells at various stages of development. Proc Natl Acad Sci USA 1994;91:10962–10966.
17 Kanzler H, Küppers R, Hansmann ML, Rajewsky K: Hodgkin and Reed-Sternberg cells in Hodgkin's disease represent the outgrowth of a dominant tumor clone derived from (crippled) germinal center B cells. J Exp Med 1996;184: 1495–1505.
18 Müschen M, Rajewsky K, Braeuninger A, Baur AS, Oudejans JJ, Roers A, Hansmann ML, Kueppers R: Rare occurrence of classical Hodgkin's disease as a T-cell lymphoma. J Exp Med 2000;191:387–394.
19 Luo L, Salunga RC, Guo H, Bittner A, Joy KC, Galindo JE, Xiao H, Rogers KE, Wan JS, Jackson MR, Erlander MG: Gene expression profiles of laser-captured adjacent neuronal subtypes. Nat Med 1999;5:117–122.
20 Jin L, Thompson CA, Qian X, Kuecker SJ, Kulig E, Lloyd RV: Analysis of anterior pituitary hormone mRNA expression in immunophenotypically characterized single cells after laser capture microdissection. Lab Invest 1999; 79:511–512.
21 Quintanilla-Martinez L, Fend F, Rodriguez Moguel L, Spilove L, Beaty MW, Kingma DW, Raffeld M, Jaffe ES: Peripheral T-cell lymphoma with Reed-Sternberg-like cells of B-cell phenotype and genotype associated with Epstein-Barr virus infection. Am J Surg Pathol 1999; 23:1233–1240.
22 Kremer M, Sandherr M, Geist B, Cabras AD, Höfler H, Fend F: EBV-negative Hodgkin's lymphoma following mycosis fungoides: Evidence for distinct clonal origin. Mod Pathol, in press.
23 Suarez-Quian CA, Goldstein SR, Bonner RF: Laser capture microdissection: A new tool for the study of spermatogenesis. J Androl 2000; 21:601–608.
24 Dietmaier W, Hartmann A, Wallinger S, Heinmoller E, Kerner T, Endl E, Jauch KW, Hofstädter F, Rüschoff J: Multiple mutation analyses in single tumor cells with improved whole genome amplification. Am J Pathol 1999;154: 83–95.
25 Shen CY, Yu JC, Lo YL, Kuo CH, Yue CT, Jou YS, Huang CS, Lung JC, Wu CW: Genome-wide search for loss of heterozygosity using laser capture microdissected tissue of breast carcinoma: An implication for mutator phenotype and breast cancer pathogenesis. Cancer Res 2000;60:3884–3892.

26 Fink L, Seeger W, Ermert L, Hänze J, Stahl U, Grimminger F, Kummer W, Bohle RM: Real-time quantitative RT-PCR after laser-assisted cell picking. Nat Med 1998;4:1329–1333.
27 Goldsworthy SM, Stockton PS, Trempus CS, Foley JF, Maronpot RR: Effects of fixation on RNA extraction and amplification from laser capture microdissected tissue. Mol Carcinog 1999;25:86–91.
28 Hiller T, Snell L, Watson P: Microdissection RT-PCR analysis of gene expression in pathologically defined frozen tissues. Biotechniques 1996;21:38–44.
29 Kohda Y, Murakami H, Moe OW, Star RA: Analysis of segmental renal gene expression by laser capture microdissection. Kidney Int 2000;57:321–331.
30 Krizman DB, Chuaqui RF, Meltzer PS, Trent JM, Duray PH, Lineham WM, Liotta LA, Emmert-Buck MR: Construction of a representative cDNA library from prostatic intraepithelial neoplasia. Cancer Res 1996;56:5380–5383.
31 Leethanakul C, Patel V, Gillespie J, Shillitoe E, Kellman RM, Ensley JF, Limwongse V, Emmert-Buck MR, Krizman DB, Gutkind JS: Gene expression profiles in squamous cell carcinomas of the oral cavity: Use of laser capture microdissection for the construction and analysis of stage-specific cDNA libraries. Oral Oncol 2000;36:474–483.
32 Leethanakul C, Patel V, Gillespie J, Pallente M, Ensley JF, Koontongkaew S, Liotta LA, Emmert-Buck M, Gutkind JS: Distinct pattern of expression of differentiation and growth-related genes in squamous cell carcinomas of the head and neck revealed by the use of laser capture microdissection and cDNA arrays. Oncogene 2000;19:3220–3224.
33 Maitra A, Wistuba II, Virmani AK, Sakaguchi M, Park I, Stucky A, Milchgrub S, Gibbons D, Minna JD, Gazdar AF: Enrichment of epithelial cells for molecular studies. Nat Med 1999; 5:459–463.
34 Peterson LA, Brown MR, Carlisle AJ, Kohn EC, Liotta LA, Emmert-Buck MR, Krizman DB: An improved method for construction of directionally cloned cDNA libraries from microdissected cells. Cancer Res 1998;58:5326–5328.
35 Sgroi DC, Teng S, Robinson G, LeVangie R, Hudson JR Jr, Elkahloun AG: In vivo gene expression profile analysis of human breast cancer progression. Cancer Res 1999;59:5656–5661.
36 Shibutani M, Uneyama C, Miyazaki K, Toyoda K, Hirose M: Methacarn fixation: A novel tool for analysis of gene expressions in paraffin-embedded tissue specimens. Lab Invest 2000; 80:199–208.
37 To MD, Done SJ, Redston M, Andrulis IL: Analysis of mRNA from microdissected frozen tissue sections without RNA isolation. Am J Pathol 1998;153:47–51.

38 Ohyama H, Zhang X, Kohno Y, Alevizos I, Posner M, Wong DT, Todd R: Laser capture microdissection-generated target sample for high-density oligonucleotide array hybridization. Biotechniques 2000;29:530–536.

39 Specht K, Richter T, Müller U, Walch A, Werner M, Höfler H: Quantitative gene expression analysis in microdissected archival formalin-fixed, paraffin-embedded tumor tissue. Am J Pathol 2000, in press.

40 Fend F, Emmert-Buck MR, Chuaqui R, Cole K, Lee J, Liotta LA, Raffeld M: Immuno-LCM: Laser capture microdissection of immunostained frozen sections for mRNA analysis. Am J Pathol 1999;154:61–66.

41 Murakami H, Liotta L, Star RA: IF-LCM: Laser capture microdissection of immunofluorescently defined cells for mRNA analysis. Kidney Int 2000;58:1346–1353.

42 Walch AK, Zitzelsberger HF, Bruch J, Keller G, Angermeier D, Aubele MM, Mueller J, Stein H, Braselmann H, Siewert JR, Hofler H, Werner M: Chromosomal imbalances in Barrett's adenocarcinoma and the metaplasia-dysplasia-carcinoma sequence. Am J Pathol 2000; 156:555–566.

43 Banks RE, Dunn MJ, Forbes MA, Stanley A, Pappin D, Naven T, Gough M, Harnden P, Selby PJ: The potential use of laser capture microdissection to selectively obtain distinct populations of cells for proteomic analysis – Preliminary findings. Electrophoresis 1999;20: 689–700.

44 Emmert-Buck MR, Gillespie JW, Paweletz CP, Ornstein DK, Basrur V, Appella E, Wang QH, Huang J, Hu N, Taylor P, Petricoin EF 3rd: An approach to proteomic analysis of human tumors. Mol Carcinog 2000;27:158–165.

45 Natkunam Y, Rouse RV, Zhu S, Fisher C, van De Rijn M: Immunoblot analysis of CD34 expression in histologically diverse neoplasms. Am J Pathol 2000;156:21–27.

46 Ornstein DK, Englert C, Gillespie JW, Paweletz CP, Linehan WM, Emmert-Buck MR, Petricoin EF 3rd: Characterization of intracellular prostate-specific antigen from laser capture microdissected benign and malignant prostatic epithelium. Clin Cancer Res 2000;6:353–356.

47 Ornstein DK, Gillespie JW, Paweletz CP, Duray PH, Herring J, Vocke CD, Topalian SL, Bostwick DG, Linehan WM, Petricoin EF 3rd, Emmert-Buck MR: Proteomic analysis of laser capture microdissected human prostate cancer and in vitro prostate cell lines. Electrophoresis 2000;21:2235–2242.

48 Simone NL, Remaley AT, Charboneau L, Petricoin EF 3rd, Glickman JW, Emmert-Buck MR, Fleisher TA, Liotta LA: Sensitive immunoassay of tissue cell proteins procured by laser capture microdissection. Am J Pathol 2000; 156:445–452.

Recovering DNA and Optimizing PCR Conditions from Microdissected Formalin-Fixed and Paraffin-Embedded Materials

Zhi-Ping Ren   Jan Sällström   Christer Sundström   Monica Nistér
Yngve Olsson

Department of Genetics and Pathology, Rudbeck Laboratory, Unversity Hospital, Uppsala, Sweden

Key Words

Microdissection · Formalin-fixed material · DNA · PCR

Abstract

Microdissection is a powerful technique in molecular pathology and genetic investigations. To detect genetic alterations such as gene mutation or deletion from tumor specimen, the purity of target cells is extremely critical. Unwanted cell contamination will dramatically dilute the detectable level of the abnormality. The major obstacle in clinical research is to obtain sufficient and qualified DNA from a small amount of formalin-fixed and paraffin-embedded materials. We have successfully modified our previous protocols and overcome the difficulties of recovery of DNA. After these modifications, almost every single formalin-fixed and paraffin-embedded specimen has been successfully amplified in the required DNA region.

Copyright © 2001 S. Karger AG, Basel

Introduction

The microdissection technique has become more popular because of its accuracy in picking up specific target cells in heterogeneous cell populations. It is essential to obtain pure target cells for detecting gene alterations in molecular and genetic analysis. Microdissection combined with other sensitive methods, such as PCR and DNA sequencing, will allow us to correlate genetic alterations and changes in morphologically defined cell populations. In the future, the microarray technology may well play an important role in molecular diagnosis and prognosis. The microarray-detectable levels of the genetic abnormalities also depend on the purity of the cell populations which have been obtained from the samples.

Unwanted cell contamination will dramatically reduce the detection level of the genetic alterations. For instance, if one normal lymphocyte is mixed with one tumor cell with p53 gene mutation in one of the two alleles, the p53 gene mutation signal is only up to 25%. It is, therefore, extremely important to obtain target cell samples as pure as possible for a molecular genetic analysis. Instead of the manual microdissection, there are at least four different types of microdissection equipment which can be used even at a single cell level [1–3].

The most difficult part in clinical research for genetic analysis is to get sufficient and qualified DNA from formalin-fixed and paraffin-embedded materials. This is particularly true when the sample size is very small, such as around 50 cells. The problem of recovering DNA is caused by influences from the fixatives. There are two different kinds of fixatives, cross-linking and precipitation fixatives. Precipitation fixatives, such as ethanol, methanol and acetone, preserve molecular and cellular substances but the morphology is relatively poor. Cross-linking fixatives, such as formaldehyde, glutaraldehyde and paraformaldehyde, preserve morphology but have comparatively poor molecular preservation and recovery. Formalin is the most commonly used cross-linking fixative. All archive materials used in our clinical research are derived from formalin-fixed and paraffin-embedded tissue blocks.

DNA of all chromosomes is packed into a compact structure with the aid of specialized proteins, histones and nonhistone chromosomal proteins. Chromosomes are composed of histone and nonhistone proteins, DNA and a small amount of RNA. Histones are the principal structural proteins of the eucaryotic chromosome [4–6]. Histones are positively charged amino acids. The positive charge helps the histones to be tightly bound to DNA which is highly negatively charged. The DNA double helix coils the histones and folds many times in a special way to a compact structure. If it were stretched out, the DNA double helix in each human chromosome would span the cell nucleus thousands of times [7, 8]. The formalin fixative penetrates the tissue at the rate of 0.5 mm/h [9] and cross-links all the proteins in the chromosomes. The longer the fixation time is, the stronger cross-linking will become. Formalin causes a tight, stable cross-linking network in the tissue, preserves the entire cellular structures and prevents DNA double helix release from the complex. If the tissue is fixed for too long, the DNA double helix can be broken down because of the very strong cross-linking effect.

The key to a successful DNA preparation is to completely digest all the proteins in the tissue samples. Any proteins leftover in the chromosomes will severely interrupt the quality as well as the quantity of the DNA template. Based on our previous protocol [10–12], we have now modified our methods regarding DNA preparation and optimized the PCR condition. With the new modification, virtually all of our investigated formalin-fixed paraffin-embedded microdissected samples except one (99%, n = 111) showed a successful recovery of DNA. The size of the microdissected samples was from 50 to 6,000 cells. They originated from cancers of the urinary bladder and gliomas.

DNA Preparation

Briefly, formalin-fixed paraffin-embedded samples were sectioned to 6 μm on plain glass slides. A consecutive section was stained with HE for visualizing target cells and served as a control for confirming the accuracy of the microdissection and the precise morphological identification. Slides were deparaffinized and stained by HE or immunostained prior to microdissection without coverslip. Different microdissected areas were carefully isolated manually and picked up by a small scalpel under light microscopy. When the samples were very difficult to microdissect by hand, the PALM system was applied to ensure the purity and transfer. Samples were transferred to a tube containing 100 μl proteinase K buffer (10 mM Tris-HCl, pH = 8.0, 1 mM EDTA and 1% Tween-20). Cells were lysed at 56°C overnight or during several days by adding 2–8 μl proteinase K (20 μg/μl) to the proteinase K buffer. Samples were centrifuged at 14,000 rpm for 10 min. The supernatant of the samples was transferred to a new tube. After incubation at 95°C for 10 min to inactivate the proteinase K, the sample DNA was ready for PCR.

Proteinase K is used in DNA preparation. The amount of proteinase K in each sample can differ depending on the amount of material and the quality of the material. If the material has been fixed for a long period of time or a lot of microdissected material is collected, more proteinase K is needed in the DNA extraction. The concentration of proteinase K is 0.04 up to 0.16%. In the majority of cases, 0.04% of proteinase K is enough to digest the proteins in the microdissected samples completely. A further increase of proteinase K concentration above 0.16% is not necessary. A conventional proteinase K buffer can be used (10 mM Tris-HCl, pH = 8.0, 1 mM EDTA, 1% Tween-20). Tween-20 is a detergent used for breaking the cell membranes. However, it may inhibit the PCR reaction when the concentration is increased to a certain level. In some cases where cell membranes and nuclear membranes are not intact, Tween-20 may be subtracted from the buffer. The alternative is to use the PCR buffer (10 mM Tris-HCl, pH = 8.3, 50 mM KCl) instead of proteinase K buffer, which works fine in our hands. The temperature for protein digestion and the time of digestion can be individually adjusted. Proteinase K works best at 56°C and can be degraded for a period of time. Overnight proteinase K digestion normally is enough in the majority of cases. If the buffer is not clear and some debris can still be observed additional proteinase K and a longer incubation time are definitely required. This procedure can be repeated for several days until all the proteins is digested and the buffer becomes clear. Proteins are the inhibitor of PCR; in case of any incompletely digested protein or other insoluble substances left in the buffer, samples should be centrifuged at high speed and only the supernatant should be collected as DNA template for PCR.

Optimizing PCR Conditions

There are plenty of books and PCR application manuals explaining how to optimize PCR conditions. All the principles of PCR are applicable in formalin-fixed, paraffin-embedded and microdissected samples.

This paper looks into optimization of PCR in microdissected and formalin-fixed materials. A few points should be emphasized regarding microdissected samples particularly the amount of DNA template and the quality of the DNA. The copy numbers of sample DNA from microdissected materials are fairly small. They are too small to even measure the DNA concentration of the samples.

Recovering completely released DNA from cross-linked proteins is the most important part in DNA preparation. If the DNA template is pure, optimizing PCR conditions will be much easier. Annealing temperature, magnesium concentration, choice of DNA polymerase, cycle number, primer design and concentration are important factors for successful PCR.

For formalin-fixed microdissected materials, it is sometimes difficult to amplify the gene by using the primers which work very well in cell lines or fresh frozen materials. It is probably due to the fact that the DNA double helix has been broken down into smaller fragments. Therefore, a new set of primer pairs which amplify a shorter fragment, such as around 120 base pairs, is needed. The PCR cycle number should increase to 40 or even to 50 cycles or a nested PCR is required in few very difficult cases because there is so little DNA template in the microdissected material. Of course, we are faced with the risk of false-positive PCR. If we increase the PCR cycle number, we also increase the noise level of the PCR. Our strategy is to get the positive result first and then try to develop a strict confirmation system to eliminate the possibility of false-positive results.

All the registered mutations in our clinical studies were confirmed a second time by repeating the DNA sequence analysis of the original extracts from the formalin-fixed microdissected materials. In some instances the entire procedure including microdissection was repeated to ensure the accuracy of the analysis.

Acknowledgment

This work was supported by the Swedish Cancer Society (Cancerfonden) and the Lions Cancer Research Foundation (Lions Cancerforskningsfond).

References

1 Schutze K, Lahr G: Identification of expressed genes by laser-mediated manipulation of single cells. Nat Biotechnol 1998;16:737–742.
2 Schermelleh L, Thalhammer S, Heckl W, Posl H, Cremer T, Schutze K, Cremer M: Laser microdissection and laser pressure catapulting for the generation of chromosome-specific pain probes. Biotechniques 1999;27:362–367.
3 Hahn S, Zhong XY, Troeger C, Burgemeister R, Gloning K, Holzgreve W: Current applications of single-cell PCR. Cell Mol Life Sci 2000; 57:96–105.
4 Grunstein M: Histones as regulators of genes. Sci Am 1992;267/4:68–74B.
5 Smith MM: Histone structure and function. Curr Opin Cell Biol 1991;3:429–437.
6 Wells DE: Compilation analysis of histones and histone genes. Nucleic Acids Res 1986; 14(suppl):r119–r149.
7 Kornberg RD, Klug A: The nucleosome. Sci Am 1981;244/2:52–64.
8 Richmond TJ, Finch JT, Rushton B, Rhodes D, Klug A: Structure of the nucleosome core particle at 7 A resolution. Nature 1984;311: 532–537.
9 Damjanov I, Lindner J: Anderson's Pathology, ed 10. St. Louis, Mosby-Year Book, 1996, p 121.
10 Hedrum A, Pontén F, Ren ZP, Lundeberg J, Pontén J, Uhlén M: Sequence-based analysis of the human p53 gene based on microdissection of tumor biopsy samples. Biotechniques 1994; 17/1:118–119, 122–124, 126–129.
11 Ren ZP, Hedrum A, Pontén F, Nistér M, Ahmadian A, Lundeberg J, Uhlén M, Pontén J: Human epidermal cancer and accompanying precursors have identical p53 mutations different from p53 mutations in adjacent areas of clonally expanded non-neoplastic keratinocytes. Oncogene 1996;12:765–773.
12 Ren ZP, Ahmadian A, Pontén F, Nistér M, Berg C, Lundeberg J, Uhlén M, Pontén J: Benign clonal keratinocyte patches with p53 mutations show no genetic link to synchronous squamous cell precancer or cancer in human skin. Am J Pathol 1997;150:1791–1803.

Diagnosis of Papillary Thyroid Carcinoma Is Facilitated by Using an RT-PCR Approach on Laser-Microdissected Archival Material to Detect RET Oncogene Activation

G. Lahr[a] M. Stich[a] K. Schütze[a] P. Blümel[b] H. Pösl[a] W.B.J. Nathrath[b]

[a]Laser Laboratory and Molecular Biology, First Medical Department and [b]Department of Pathology, Municipal and Teaching Hospital München-Harlaching, Munich, Germany

Key Words
RET/PTC1 · Papillary thyroid carcinoma · Laser microdissection · RT-PCR · Laser microbeam microdissection · Laser pressure catapulting · Tissue, archival

Abstract

Objective: The purpose of this study was to investigate the value of the expression of the RET oncogene (rearranged during transfection) in papillary thyroid carcinomas (PTC) and its variants in the differential diagnosis of thyroid neoplasias. According to the literature RET oncogene activation by chromosomal rearrangements has been exclusively implicated in PTCs. ***Methods:*** To establish the incidence of RET activation in PTCs we used 5- to 10-μm sections from archival paraffin blocks. Either parts of the tissue slices were manually dissected or a few distinct cells were microdissected by laser-mediated manipulation with the Robot-MicroBeam system. RNA was extracted from paraffin-embedded thyroid tumors and the corresponding normal tissue. RT and nested PCR were performed using primers for RET/PTC1, PTC2 and PTC3, or for RET exons 12 and 13. PCR products were resolved by gel electrophoresis. ***Results:*** We detected RET transcription in approximately 85% of the PTCs including follicular variants and in isolated cells of the same tissues, but not in nonmalignant thyroid tissue. ***Conclusions:*** Our method may serve as an additional diagnostic tool to characterize ambiguous neoplasias and to identify especially nonpapillary, i.e. follicular tumors, as papillary carcinomas. Additionally, this study has demonstrated that expressed genes can be analyzed from routine histopathological tissue slides or pooled single cells. Large retrospective studies can also be performed with this method.

Copyright © 2001 S. Karger AG, Basel

Introduction

There are two major categories of malignant thyroid follicular cell neoplasms, the papillary and the follicular carcinoma. Papillary thyroid carcinoma (PTC) is the most common type of thyroid tumor, representing about 70% of all thyroid carcinomas. PTCs usually carry a good prognosis (90–95% survival at 5 years), despite often presenting lymph node metastases at diagnosis [1, 2]. In accordance with the distinction of the above two thyroid carcinoma types, two pathways of the follicular cell carcinogenesis can be postulated: the papillary and the follicular thyroid carcinoma pathway [3, 4]. While RAS mutations are present to a high percentage in follicular carcinomas [5, 6], the papillary pathway is mostly defined by a rearranged proto-oncogene named RET which fuses to unrelated genes (RET/PTC). By this translocation the tyrosine kinase (TK) domain is fused to a variety of other 5′ elements and the RET oncogene is activated. The RET proto-oncogene encodes a cell surface glycoprotein related to the family of receptor TK [7]. RET is normally expressed in neural crest-derived tissues including the interstitial C cells of the thyroid and the medullary carcinomas. The RET proto-oncogene expression has not been detected in normal follicular cells, the cells of origin of papillary carcinoma [8]. Rearrangements of the RET proto-oncogene have only been found in vivo in thyroid gland tumors of the papillary histotype, thus representing a fascinating example of organ and tumor specificity [9, 10]. The rearranged RET/PTCs are mainly restricted to PTC; only a very small percentage of RET/PTCs has been reported in possible follicular-type lesions [9, 11, 12]. Three main forms, PTC1, PTC2 and PTC3, are the most common; in addition, two variations of RET/PTC3 have been described: RET/PTC4 and RET/PTC5 [13, 14]. RET/PTC1 is formed by an intrachromosomal rearrangement of chromosome 10, fusing the RET TK domain to the 5′terminal region of the H4 gene [15, 16]. The frequency of RET/PTC1 occurrence in PTCs varies in different populations between 25 and 60% [12]. RET/PTC2 is characterized by involvement of an RI alpha regulatory subunit of protein kinase A localized on chromosome 17 [17]. RET/PTC3 is generated by an intrachromosomal rearrangement with the gene RFG/ELE1 [18]. The breakpoint of each of these three RET rearrangements occurs in intron 11, immediately upstream of the sequence that codes for the RET TK receptor.

RET/PTC analysis is most commonly performed using the RT-PCR technique. This approach can also be used for archival tissue. Formalin-fixed, paraffin-embedded tissue is an invaluable resource for molecular genetic studies. It was shown that neither histological tissue preparation methods like fixation nor staining (including immunostaining) do negatively impact RNA recovery [19–21]. The ability to study preserved tissues at the molecular level makes possible retrospective studies on large numbers of patients and may permit tracking, over long periods of time, of genetic changes that are associated with diseases.

The major categories of malignant thyroid follicular cell neoplasms, the papillary and the follicular carcinoma, have been distinguished on the basis of their architectural features. However, a number of PTCs exhibit follicular growth patterns, and in some cases the distinction between the follicular variant of the papillary carcinoma, a follicular adenoma, represents a diagnostic dilemma. Since the clinical and biological behavior of the follicular variants matches that of the conventional papillary carcinomas, and since these in general differ from the follicular carcinomas not only clinically but also in their genetic setup, the identification of a distinctive genetic marker like RET/PTC could be of help in the differential diagnosis between papillary and follicular neoplasms of the thyroid.

Materials and Methods

Tissues

24 different thyroid tissues from 13 patients (7 female, 6 male, mean age 45 years) operated in for goiter were included in this study and consisted of 16 papillary carcinomas, 1 follicular carcinoma, 1 atypical oncocytic adenoma, 5 adenomatoid/colloid goiter lesions only, and 1 ambiguous follicular neoplasm, in which the diagnosis of a carcinoma seemed not sufficiently justified by histopathological criteria (tissue No. 12; table 1).

Specimen Preparation

5- to 10-μm serial sections of routinely formalin-fixed and paraffin-embedded thyroid tumors and adjacent normal tissue were prepared. The sections were deparaffinized and consecutive sections were stained with hematoxylin-eosin as usual for microscopic examination. For a better visualization during laser-assisted microdissection the selected tissues were stained for 1 min with Mayer's hematoxylin solution (Sigma, Deisenhofen, Germany) and rinsed in H_2O for 30 s and dried.

Immunohistochemistry

Interstitial C cells, a potential source of RET proto-oncogene expression, were immunostained with a polyclonal antibody for calcitonin (Cal5F5, Dakopatts, Hamburg, Germany). 5- to 10-μm sections were cut and mounted on superfrosted slides. After deparaffinization the sections were equilibrated for 5 min in Tris-buffered saline (TBS), blocked with a serum blocking solution (Histostain-PLUS Bulk Kit, Zymed, Berlin, Germany) for 10 min followed by incubation for 60 min with a 1:50 dilution of the calcitonin antibody. After washing the sections were incubated with the biotinylated second antibody reagent (Histostain-PLUS Bulk Kit) for 10 min. After washing in TBS the sections were incubated for 10 min with the enzyme conjugate and washed in TBS. Subsequently, sections were stained with AEC (Histostain-PLUS Bulk Kit) for 5 min, washed in TBS and air-dried.

Laser Microdissection with the Robot-MicroBeam

The Robot-MicroBeam (PALM, Bernried, Germany) consists of a pulsed, low energy 337-nm nitrogen laser with high beam quality and an inverted microscope (Axiovert 135, Carl Zeiss, Göttingen, Germany). The laser microbeam was focused through a 20× or 40× dry long distance objective lens. Unwanted tumor cells as well as immunohistochemically stained C cells were ablated by the laser light. For microdissection of follicular epithelial tumor cells from dif-

Table 1. Summary of the specimen specifications of the tissues investigated and corresponding tissue numbers

Tissue No.	Localization	Diagnosis	Gender/age[1]	RET/PTC1 PCR	RET/PTC2 PCR	RET/PTC3 PCR	IHC Calc	β-Actin PCR
1	lymph node, left neck	lymph node metastasis of a papillary carcinoma	female/59 (1)	–	–	–	–	+
2	thyroid left lobe	follicular carcinoma; colloid goiter	male/62 (2)	–	–	–	–	–
3	thyroid left lobe	hyperplastic nodule of adenomatoid goiter	female/59 (1)	–	–	–	n.d.	+
4	thyroid right lobe and isthmus	hyperplastic nodule of adenomatoid goiter	female/59 (1)	–	–	–	–	+
5	thyroid left lobe	papillary carcinoma, follicular variant; adenomatoid goiter	female/59 (1)	–	–	–	–	+
6	thyroid right lobe and isthmus	papillary microcarcinoma; adenomatoid goiter	female/59 (1)	+	–	–	–	+
7	thyroid right lobe	papillary carcinoma	male/32 (3)	+	–	–	+	+
8	thyroid left lobe	papillary carcinoma, follicular tall cell variant; adenomatoid goiter	female/52 (4)	+	–	–	n.d.	+
9	thyroid left lobe	papillary carcinoma, follicular variant; adenomatoid goiter	male/49 (5)	+	–	–	+	+
10	thyroid left lobe and isthmus	papillary carcinoma, follicular tall cell variant; Hashimoto thyroiditis	female/27 (6)	+	–	–	–	+
11	thyroid left lobe and isthmus	papillary carcinoma, follicular tall cell variant; Hashimoto thyroiditis	female/27 (6)	+	–	–	–	+
12	thyroid left lobe	adenomatoid goiter; non-Hashimoto thyroiditis; no carcinoma	female/65 (7)	+	–	–	n.d.	+
13	thyroid right lobe and isthmus	papillary carcinoma, follicular variant; adenomatoid goiter; thyroiditis	female/65 (7)	+	–	–	–	+
14	thyroid right lobe	papillary microcarcinoma; adenomatoid goiter	male/51 (8)	+	–	–	–	+
15	thyroid right lobe	papillary carcinoma; adenomatoid goiter	male/50 (9)	+	–	–	–	+
16	thyroid left lobe	adenomatoid goiter	male/50 (9)	–	–	–	n.d.	+
17	thyroid left lobe	papillary carcinoma; colloid goiter	male/69 (10)	+	–	–	n.d.	+
18	thyroid right lobe and isthmus	papillary carcinoma; adenomatoid goiter; thyroiditis	female/65 (7)	+	–	–	–	+
19	thyroid right lobe	hyperplastic nodule of adenomatoid goiter	female/46 (11)	–	–	–	–	–
20	thyroid right lobe	colloid goiter	male/32 (3)	–	–	–	–	+
21	thyroid left lobe	papillary carcinoma, follicular variant; adenomatoid goiter	male/50 (9)	+	–	–	n.d.	+
22	thyroid right lobe	papillary carcinoma, follicular variant; adenomatoid goiter; thyroiditis	female/36 (12)	+	–	–	n.d.	+
23	thyroid left lobe	papillary carcinoma, follicular variant	female/36 (12)	+	–	–	n.d.	+
24	thyroid left lobe	struma node; atypical oncocytic adenoma	female/33 (13)	–	–	–	n.d.	+

RT-PCR results for RET/PTC1, PTC2, PTC3, and β-actin as well as results of the immunohistochemical staining (IHC) for calcitonin (Cals) are given. The same patient No. denotes tissue from the same patient who was operated on for goiter. n.d. = Not determined.

[1] Age is given in years. Figures in parentheses represent patient No.

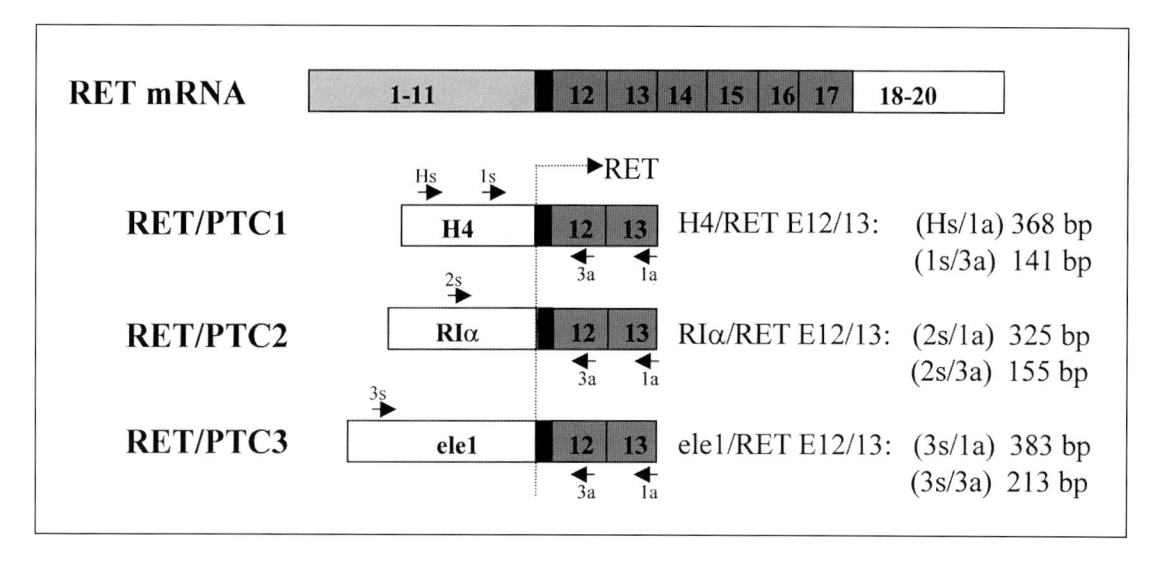

Fig. 1. Schematic illustration of RET proto-oncogene mRNA and RET/PTC1, PTC2 and PTC3 exon structure. The TK domain is shown in black and dark gray. The relative position of the outnested and nested primers is indicated as well as the size of the expected PCR products.

ferent specimens the laser energy was adjusted to catapult the cells directly from the glass slide into a common microfuge cap moistened with a 2-μl droplet of mineral oil which was held and centered above the line of laser fire by a special collector device. By this procedure about 50 pooled carcinoma cells were collected from each tissue sample. The microdissected cells were simply covered with the buffer included in the total RNA isolation kit. Then the caps were topped with the remaining tube and processed as described below.

Isolation of Total RNA

Total RNA from the thyroid tumor cell line TPC-1 carrying a RET/PTC1 rearrangement (a generous gift of Dr. B. Mayr, Med. Hochschule Hannover, Germany) as well as microdissected tumor cells were isolated with the PUREscript RNA isolation Kit (BIOzym Diagnostik, Hessisch Oldendorf, Germany) according to the manufacturers protocol. Total RNA of TPC-1 was dissolved in 50 μl and RNA of microdissected cells in 5 μl RNA hydration solution supplied by the manufacturer. The manually dissected tissue samples were dissolved in 30–200 μl proteinase K (PK) buffer depending on the amount of tissue. PK buffer contained 50 mM Tris-HCl (pH 8.3), 1 mM EDTA, 0.5% Tween-20 and 10.7 mg/ml PK. Sections were incubated at 56°C for up to 3 h or overnight at 37°C [22]. PK was inactivated at 85°C for 10 min.

RT-PCR from Tissue Sections and Microdissected Cells

cDNA synthesis was performed with the ExpeRT-PCR Kit (Hybaid-AGS, Heidelberg, Germany) using the two-step protocol in a 25-μl reaction volume. The reaction mix included 2 μl 10× random hexamers (Roche Diagnostics, Mannheim, Germany) and 3–5 μl of the PK digest from manually dissected cell clusters. We used 5 μl of total RNA isolated from laser-microdissected cells. The RT reactions were incubated for 60 min at 42°C. Nested and heminested PCR reactions were performed using primers for RET/PTC1, PTC2 and PTC3, and for RET exons 12 and 13 (fig. 1).

The RT reaction was followed by the first PCR reaction for PTC1 using up to 12.5 μl (microdissected cells) or 3–5 μl (manually dissected cells) RT reaction mix, 10 pmol downstream 1a primer and 10 pmol upstream Hs primer resulting in a 368-bp fragment. The PCR mix can optionally include additional primers for β-actin mRNA to yield a 271-bp fragment. The first PCR reaction for PTC2 detection was performed with primers 2s/1a, resulting in a 325-bp fragment. For PTC3 detection the primer pair 3s/1a was used generating a 383-bp fragment. The mixtures were subjected to 1 cycle at 94°C for 2 min, 30 cycles at 94°C for 30 s, 60°C for 30 s, 72°C for 1 min, followed by 1 cycle at 72°C for 15 min and storage at 4°C. Controls containing H$_2$O instead of DNA (K) were always run in parallel.

Nested and Heminested PCR

Nested PCR amplification was performed with a 5-μl aliquot of the first PCR product containing 10 pmol of 'inner' primer sets 1s/3a (PTC1), 2s/3a (PTC2) and 3s/3a (PTC3) resulting in 141-, 155- and 213-bp fragments, respectively. The samples were subjected to 30 cycles as mentioned above. Controls containing H$_2$O instead of DNA (K) were always run in parallel. PCR products were resolved on 1.8% agarose gels in TAE buffer containing 0.25 μg/ml ethidium bromide. The 100-bp ladder size markers (Gibco-BRL, Eggenstein, Germany) were run side by side with the generated PCR products.

The sequences for human Ret and RET/PTC primers [GenBank accession No.: Humptcaa.gb_pr1 (PTC1); Humretri.gb_pr1 (PTC2); S71225i.gb_pr1 (PTC3)] were:

RET/PTC1:
Hs: GGCATTGTCATCTCGCCGTTC (H4)
1s: CAGCAAGAGAACAAGGTGCTG (H4)
1a: CTGCTTCAGGACGTTGAACTC (exon 13 RET)
3a: CAAGTTCTTCCGAGGGAATTC (exon 12 RET)

Fig. 2. Histology of an ambiguous tissue sample typed as adenomatoid goiter, non-Hashimoto thyroiditis. No carcinoma was histologically recognizable in this section. Stars indicate equivalent positions in the tissue slice. **a** 10× objective. **b** 20× objective.

Fig. 3. RET/PTC1 RT-PCR-amplified products spanning the H4 gene and codons 12 and 13 of RET loaded on an ethidium bromide-stained agarose gel. **a** Tissue No. 1–10. **b** Tissue No. 11–20. Nested (second PCR) products of 141-bp (RET/PTC1) and 217-bp (β-actin) products are indicated (black and white arrows, respectively). Lanes 1–20 are derived from specimens (tissue No.) depicted in table 1. T = TPC-1 cell line; Md = medullary thyroid carcinoma; K = H₂O instead of cDNA; M = 100-bp ladder size marker. Specific band sizes are indicated in the middle.

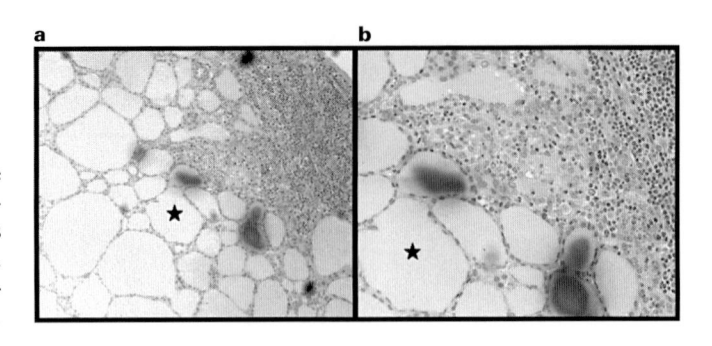

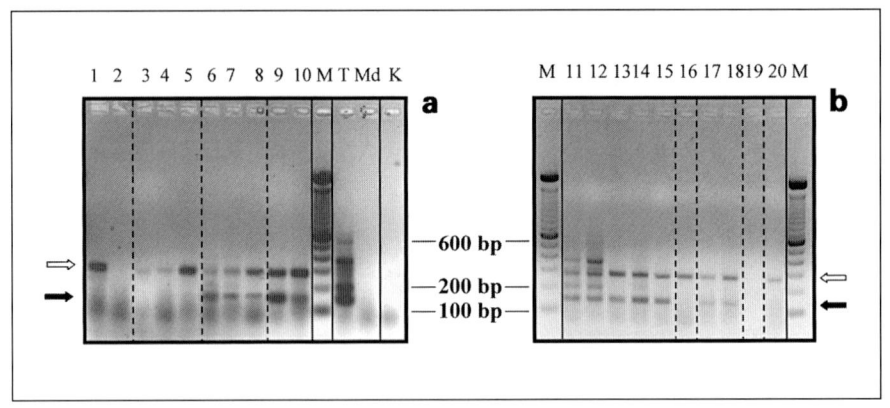

RET/PTC2 and 3:
2s: GAGGGAGCTTTGGAGAACTTG (RI alpha)
3s: GAGAAGAGAGGCTGTATCTCC (ele 1)

β-actin:
h-acts: CTA CAA TGA GCT GCG TGT GGC (exon 2)
h-acta: CAG GTC CAG ACG CAG GAT GGC (exon 3)

Results

In our study we analyzed 24 tissue samples of 13 different patients with thyroid abnormalities (see table 1). In the tissue collection we also included a tissue section from a patient suffering from thyroiditis (tissue No. 12) containing obviously no malignant cells (fig. 2).

RT-PCR of Manually Dissected Tissue Slices

In a first attempt we manually dissected larger tissue areas from the tissue slices. After PK incubation these tissue extracts were reverse-transcribed. The generated cDNA was amplified by nested PCR with primers specific for RET/PTC1, PTC2 and PTC3 and β-actin. By use of specific primers hybridizing in the upstream-fused genes and using primers in the TK domain of RET, we could determine directly RET/PTC translocations. All rearrangements were of the PTC1 type. PTC2 or PTC3 transcriptions were not found (data not shown) [23]. RET/PTC1 transcription was comparably present in both the conventional papillary carcinoma tissues (No. 6, 7, 14, 15, 17 and 18) and in the follicular variant tissues (No. 5, 8–11 and 13; fig. 3a, b). The two carcinoma lesions without an RET/PTC transcription (tissue No. 1 and 5; table 1) were the left-sided primary papillary carcinoma and its metastasis in a left neck lymph node (tissue No. 1; table 1), whereas the simultaneous papillary microcarcinoma in the right thyroid lobe of the same patient did show the RET/PTC1 transcription (tissue No. 6). The follicular carcinoma tissue (No. 2) and a medullary carcinoma, which we used as a negative control, were both negative for RET/PTC1 transcription (fig. 3a). Samples 4 and 6 originated from the same tissue block from patient 1. The papillary microcarcinoma (tissue No. 6) was dissected out manually from the area containing only adenomatoid goiter tissue. RT-PCR analysis for RET/PTC revealed that RET/PTC1 is exclusively expressed in the microcarcinoma sample. All additional adenomatoid/colloid goiter tissues (tissue No. 3, 16, 19 and 20) were also negative for RET/PTC1 as well as for RET/PTC2 and PTC3 transcription. Surprisingly, the preparation from one tissue section in one of these patients suffering from thyroidits (tissue No. 12) showed RET/PTC1 transcription (fig. 3b), although no carcinoma was histologically recognizable in this section (fig. 2).

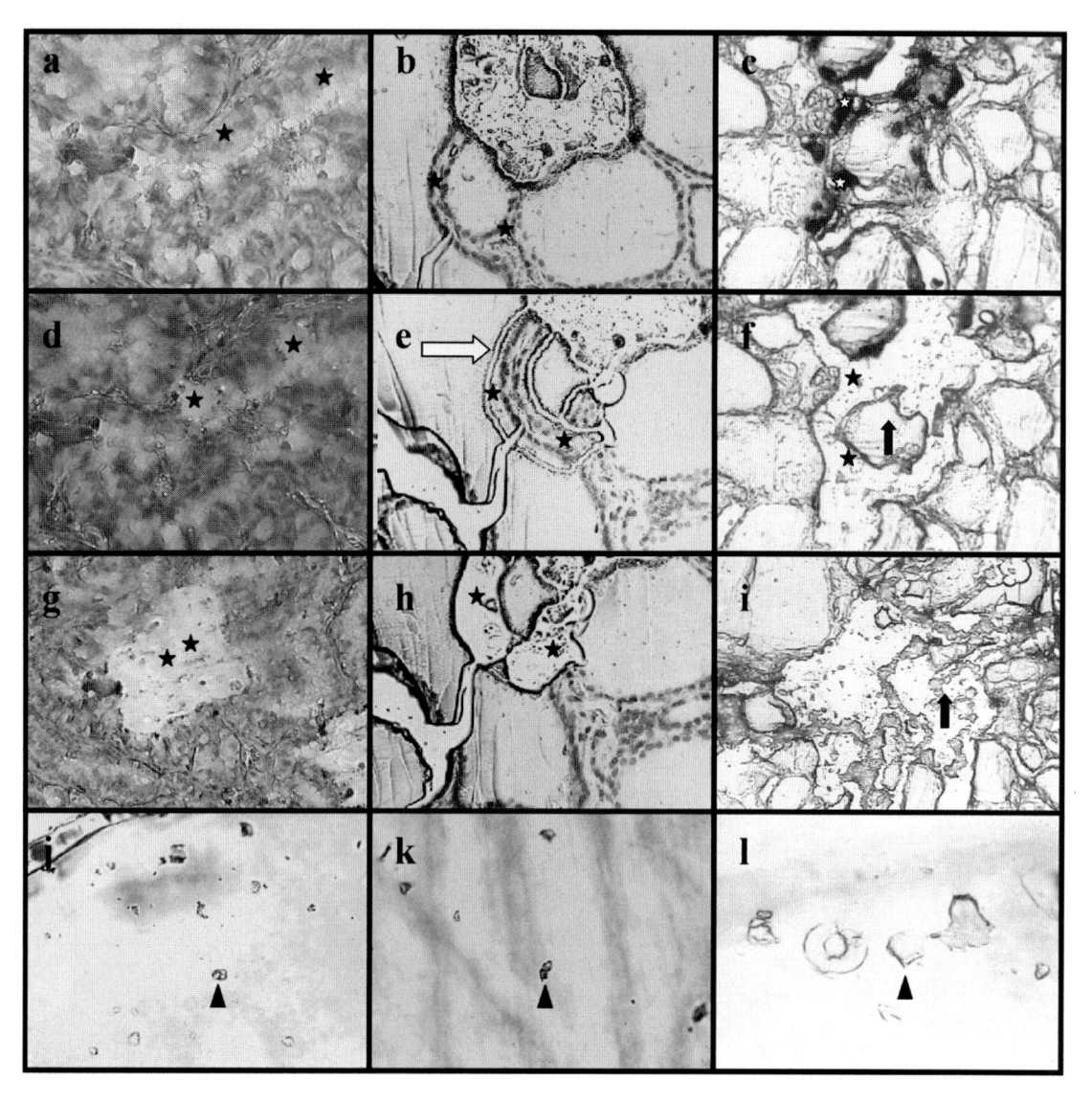

Fig. 4. Microscopic illustration of three different experiments from tissue sample No. 7, 21 and 9 (left to right) using LMM and LPC to capture cell clusters from archival hematoxylin-stained tissue slices of different PTCs. The view shows specimens before (**a–c**), after microdissection (**e**) leaving a micron-sized gap (white arrow), the remaining tissue after LPC (**g–i**) and the catapulted clusters of about 50 pooled tumor cells in the cap (**j–l**). The laser beam was also used to destroy specific papillary carcinoma cells (stars in **d**) and to destroy calcitonin-immunostained single cells (stars in **f**) within a papillary carcinoma. Stars and black arrows indicate equivalent positions. **a–f, h, i, l** 40 × objective. **g, j, k** 20 × objective.

RT-PCR of Laser-Microdissected Cells

For a more precise analysis we isolated specific tumor cells by laser-assisted microdissection (fig. 4). About 50 tumor cells each were catapulted directly from the glass slide into common microfuge caps moistened with a 2-μl droplet of mineral oil (fig. 4j–l), which were held and centered above the line of laser fire by a special collector device. In cases where the selected specimen area contained undesired cells, these could be eliminated by a direct laser shot (fig. 4d, f). Mutations in the RET oncogene are known to lead to C cell hyperplasia, and there is a theoretical possibility that C cells would account for RET TK positivity but to our knowledge not for RET/PTC translocations in the tumors examined. However, we carried out immunocytochemistry for calcitonin to identify C cells in the sections immediately adjacent to the area

Fig. 5. a, b RET/PTC1 RT-PCR-amplified products loaded on an ethidium bromide-stained agarose gel after laser microdissection of about 50 tumor cells. The nested (second PCR) product of 141 bp is spanning the H4 gene and codon 12 of RET. The lane No. correspond to sample/tissue No. depicted in table 1. Lanes 7, 9, 21 represent cells of tissue shown in figure 4. T = TPC-1 cell line; K = H$_2$O instead of cDNA; Md = medullary thyroid carcinoma; M = 100-bp ladder size marker. Black arrows indicate the 141-bp RET/PTC1 RT-PCR products.

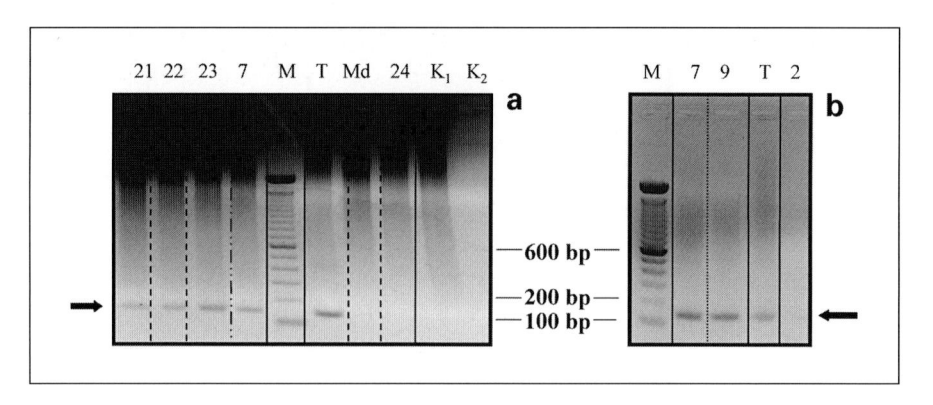

chosen for analysis. The microdissected cells were resuspended in RNA-lysis buffer. Total RNA was extracted followed by RT and RET/PTC1-specific PCR. RET/PTC1 transcription from microdissected cells was demonstrated in all 5 thyroid papillary carcinoma lesions (tissue No. 7, 9, 21–23) investigated, no matter whether C cells were ablated or not prior to microdissection. Again, RET/PTC1 transcription was comparably present in both the conventional papillary carcinoma (tissue No. 7) and in the follicular variants (tissue No. 9, 21–23) (fig. 5a, b). This translocation was absent in the medullary carcinoma, the atypical oncocytic adenoma (tissue No. 24) and in adenomatoid goiter tissue (tissue no. 20). In samples 21–23 the included C cells were not ablated.

Discussion

There is a great deal of interest in finding genetic markers that predict the behavior of thyroid tumors, as histological features alone may fail to do so. In this study we investigated the expression of rearranged RET. Thyroid follicular cells do not normally express the RET proto-oncogene [8]. However, thyroid papillary carcinoma, a tumor derived from follicular thyroid cells, shows RET activation through rearrangement, initially described with the gene H4 to result in an oncogene named RET/PTC1b [15, 16]. The fusion with H4, the most common translocation, allows the expression of the activated RET form in PTC. Here we show the detection of RET/PTC by RT-PCR analysis from cells derived from archival tissues. A previous study used the fact that RET mRNA is not normally expressed in thyroid follicular cells or PTCs, and the ability to amplify the RET TK domain in PTCs was taken as an indication of RET activation [22]. But one must be confident in using this approach that there is no C cell contamination from normal thyroid or low level 'normal' RET expression in PTC to draw this conclusion. Therefore, we used a different approach to directly test for RET/PTC positivity. We applied specific primers for the upstream-fused genes and for the TK domain of RET. In our study, all rearrangements were RET/PTC1, and were found in 85% adult PTC-positive tumors. PTC2 or PTC3 transcriptions were not found. This is in agreement with the study of Bounacer et al. [12], where specifically RET/PTC1 translocations were detected in PTC to an extent of 84%. However, the prevalence of RET/PTC rearrangements in PTC series from around the world varies widely from approximately 3 to 84% and seems to be influenced by a number of factors [12, 24–26]. Exposure to radiation and younger age of onset of PTC are each associated independently with a higher prevalence of RET/PTC [27]. In addition, it has been suggested that the presence of RET/PTC may be associated with a greater likelihood of metastatic spread and poorer prognosis [28]. In our study RET/PTC1 transcription was comparably present in both the conventional papillary carcinomas and in the follicular variants, but not in adenomatoid goiter tissues, and also not in the one follicular and in the one medullary carcinoma analyzed. Analyzing a manually microdissected papillary microcarcinoma, RT-PCR analysis for RET/PTC1 revealed that RET/PTC1 was exclusively expressed in this microcarcinoma sample. The adenomatoid goiter tissue around the microcarcinoma as well as other adenomatoid goiter tissue samples were RET/PTC1-negative. These results clearly demonstrate that the rearranged RET/PTC, especially RET/PTC1, is restricted to thyroid gland tumors of the papillary histotype, thus representing a fascinating example of tumor specificity.

RET rearrangements are usually found both in the primary tumors and in the corresponding lymph node metastases [29, 30]. The same seems to be true for RET/PTC negativity. One case of our series was negative for RET activation in the left-sided primary tumor and its metastasis in a left neck lymph node. By contrast, a simultaneous carcinoma in the right thyroid lobe of the same patient did show the RET/PTC1 transcription.

We have also shown that it is possible to extract RNA from single sections of paraffin-embedded archival material and to demonstrate gene expression by using primers specific for RET/PTC1, PTC2 and PTC3. RET/PTC1 rearrangements of papillary carcinomas were still demonstrable after the carcinoma cells had been removed by the laser-assisted microdissection technique. The presence of the RET proto-oncogene in C cells does not interfere with the demonstration of the RET/PTC rearrangement in the carcinoma cells since C cells and medullary carcinomas, the tumors derived thereof, do not express RET/PTC1.

In conclusion, RET oncogene activation by genetic rearrangements in human thyroid neoplasms is seen in the papillary cancer subtype, and RET/PTC, especially RET/PTC1 detection by RT-PCR, could be a useful marker for papillary carcinoma. If differential diagnostic problems arise in a thyroid nodule, the demonstration of RET/PTC1 in it may allow to distinguish a follicular variant of a papillary carcinoma from benign follicular lesions.

Acknowledgment

The authors want to thank H. Brand (Institute of Pathology) for excellent technical assistance in preparing the tissue sections and performing immunohistochemistry.

References

1 Hay, ID: Papillary thyroid carcinoma. Endocrinol Metab Clin North Am 1990;19:545–576.
2 Sobrinho-Simoes M: Tumours of the thyroid. A brief overview with emphasis in the most controversial issues. Curr Diagn Pathol 1995;2:15–22.
3 LiVoisi V: Surgical pathology of the thyroid; in Bennington JL (ed): Major Problems in Pathology. Philadelphia, Saunders, 1990, vol 22.
4 Rosai, J, Carcangiu ML, Delellis RA: Tumors of the thyroid gland; in Rosai J, Sobin LH (eds): Fascile 5. Washington, Armed Forces Institute of Pathology, 1992.
5 Wynford-Thomas D: Origin and progression of thyroid epithelial tumours: Cellular and molecular mechanisms. Horm Res 1997;47:145–157.
6 Lemoine NR, Mayall ES, Wyllie FS, Williams ED, Goyns M, Stringer B, Wynford-Thomas D: High frequency of ras oncogene activation in all stages of human thyroid tumorigenesis. Oncogene 1989;4:159–164.
7 Takahashi M, Buma Y, Hiai H: Isolation of ret proto-oncogene cDNA with an amino-terminal signal sequence. Oncogene 1989;4:805–806.
8 Fabien N, Paulin C, Santoro M, Berger N, Grieco M, Galvain D, Barbier Y, Dubois PM, Fusco A: Detection of RET oncogene activation in human papillary thyroid carcinomas by in situ hybridisation. Br J Cancer 1992;66:1094–1098.
9 Santoro M, Carlomagno F, Hay ID, Herrmann MA, Grieco M, Melillo R, Pierotti MA, Bongarzone I, Della Porta G, Berger N, Peix JL, Paulin C, Fabien N, Veccio G, Jenkins RB, Fusco A: Ret oncogene activation in human thyroid neoplasms is restricted to the papillary cancer subtype. J Clin Invest 1992;89:1517–1522.

10 Santoro M, Sabino N, Ishizaka Y, Ushijima T, Carlomagno F, Cerrato A, Grieco M, Battaglia C, Martelli ML, Paulin C, Fabien N, Sugimura T, Fusco A, Nagao M: Involvement of RET oncogene in human tumours: Specificity of RET activation to thyroid tumours. Br J Cancer 1993;68:460–464.
11 Ishizaka Y, Kobayashi S, Ushijima T, Hirohashi S, Sugimura T, Nagao M: Detection of retTPC/PTC transcripts in thyroid adenomas and adenomatous goiter by an RT-PCR method. Oncogene 1991;6:1667–1672.
12 Bounacer A, Wicker R, Caillou B, Cailleux AF, Sarasin A, Schlumberger M, Suarez HG: High prevalence of activating ret proto-oncogene rearrangements, in thyroid tumors from patients who had received external radiation. Oncogene 1997;15:1263–1273.
13 Fugazzola L, Pierotti MA, Vigano E, Pacini F, Vorontsova TV, Bongarzone I: Molecular and biochemical analysis of RET/PTC4, a novel oncogenic rearrangement between RET and ELE1 genes, in a post-Chernobyl papillary thyroid cancer. Oncogene 1996;13:1093–1097.
14 Klugbauer S, Demidchik EP, Lengfelder E, Rabes HM: Detection of a novel type of RET rearrangement (PTC5) in thyroid carcinomas after Chernobyl and analysis of the involved RET-fused gene RFG5. Cancer Res 1998;58:198–203.
15 Grieco M, Santoro M, Berlingieri MT, Melillo RM, Donghi R, Bongarzone I, Pierotti MA, Della Porta G, Fusco A, Vecchio G: PTC is a novel rearranged form of the ret proto-oncogene and is frequently detected in vivo in human thyroid papillary carcinomas. Cell 1990;60:557–563.

16 Pierotti MA, Santoro M, Jenkins RB, Sozzi G, Bongarzone I, Grieco M, Monzini N, Miozzo M, Herrmann MA, Fusco A, et al: Characterization of an inversion on the long arm of chromosome 10 juxtaposing D10S170 and RET and creating the oncogenic sequence RET/PTC. Proc Natl Acad Sci USA 1992;89:1616–1620.
17 Bongarzone I, Monzini N, Borrello MG, Carcano C, Ferraresi G, Arighi E, Mondellini P, Della Porta G, Pierotti MA: Molecular characterization of a thyroid tumor-specific transforming sequence formed by the fusion of ret tyrosine kinase and the regulatory subunit RI alpha of cyclic AMP-dependent protein kinase A. Mol Cell Biol 1993;13:358–366.
18 Santoro M, Wong WT, Aroca P, Santos E, Matoskova B, Grieco M, Fusco A, di Fiore PP: An epidermal growth factor receptor/ret chimera generates mitogenic and transforming signals: Evidence for a ret-specific signaling pathway. Mol Cell Biol 1994;14:663–675.
19 Schütze K, Lahr G: Identification of expressed genes by laser-mediated manipulation of single cells. Nat Biotechnol 1998;16:737–742.
20 Schütze K, Pösl H, Lahr G: Laser micromanipulation systems as universal tools in molecular biology and medicine. Mol Cell Biol 1998;44:735–746.
21 Lahr G: RT-PCR from archival single cells is a suitable method to analyze specific gene expression. Lab Invest 2000;80:1–3.
22 Williams GH, Rooney S, Thomas GA, Cummins G, Williams ED: RET activation in adult and childhood papillary thyroid carcinoma using a reverse transcriptase-n-polymerase chain reaction approach on archival-nested material. Br J Cancer 1996,74:585–589.

23 Lahr G, Stich M, Schütze K, Blümel P, Zwicknagl M, Spelsberg F, Pösl H, Nathrath WBJ: Diagnosis of papillary thyroid carcinoma is facilitated by using a RT-PCR approach on laser-microdissected archival material to detect RET oncogene activation. Virchows Arch 1999;435:293.

24 Jhiang SM, Smanik PA, Mazzaferri EL: Development of a single-step duplex RT-PCR detecting different forms of ret activation, and identification of the third form of in vivo ret activation in human papillary thyroid carcinoma. Cancer Lett 1994;78:69–76.

25 Zuo M, Shi Y, Farid NR: Low rate of ret protooncogene activation (PTC/ret/PTC) in papillary thyroid carcinomas from Saudi Arabia. Cancer 1994;73:176–180.

26 Bongarzone I, Fugazzola L, Vigneri P, Mariani L, Mondellini P, Pacini F, Basolo F, Pinchera A, Pilotti S, Pierotti MA: Age-related activation of the tyrosine kinase receptor protooncogenes RET and NTRK1 in papillary thyroid carcinoma. J Clin Endocrinol Metab 1996;81: 2006–2009.

27 Fugazzola L, Pilotti S, Pinchera A, Vorontsova TV, Mondellini P, Bongarzone I, Greco A, Astakhova L, Butti MG, Demidchik EP, et al: Oncogenic rearrangements of the RET protooncogene in papillary thyroid carcinomas from children exposed to the Chernobyl nuclear accident. Cancer Res 1995;55:5617–5620.

28 Klugbauer S, Lengfelder E, Demidchik EP, Rabes HM: High prevalence of RET rearrangement in thyroid tumors of children from Belarus after the Chernobyl reactor accident. Oncogene 1995;11:2459–2467.

29 Fusco A, Grieco M, Santoro M, Berlingieri MT, Pilotti S, Pierotti MA, Della Porta G, Vecchio G: A new oncogene in human thyroid papillary carcinomas and their lymph- nodal metastases. Nature 1987;328:170–172.

30 Bongarzone I, Butti MG, Coronelli S, Borrello MG, Santoro M, Mondellini P, Pilotti S, Fusco A, Della Porta G, Pierotti MA: Frequent activation of ret protooncogene by fusion with a new activating gene in papillary thyroid carcinomas. Cancer Res 1994;54:2979–2985.

Microsatellite Instability in Tumor and Nonneoplastic Colorectal Cells from Hereditary Non-Polyposis Colorectal Cancer and Sporadic High Microsatellite-Instable Tumor Patients

Wolfgang Dietmaier[a]   Susanne Gänsbauer[b]   Kurt Beyser[d]   Birgit Renke[c]
Arndt Hartmann[b]   Petra Rümmele[b]   Karl-Walter Jauch[c]
Ferdinand Hofstädter[b]   Josef Rüschoff[d]

[a]Molecular Pathology Diagnostic Unit, [b]Institute of Pathology, and [c]Department of Surgery, University of Regensburg, and [d]Klinikum Kassel, Germany

Key Words
Microsatellite instability · Colorectal cancer · Hereditary non-polyposis colorectal cancer · Mismatch repair · hMSH2 · hMLH1 · Immunohistochemistry

Abstract

Genetic alterations such as loss of heterozygosity (LOH) and microsatellite instability (MSI) have been frequently studied in various tumor types. Genetic heterogeneity of nonneoplastic cells has not yet been sufficiently investigated. However, genomic instability in normal cells could be a potentially important issue, in particular when these cells are used as reference in LOH and MSI analyses of tumor samples. In order to investigate possible genetic abnormalities in normal colorectal cells of tumor patients, MSI analyses of normal colonic mucosa were performed. Up to 15 different laser-microdissected normal regions containing 50–150 cells were investigated in each of 15 individual microsatellite-stable, sporadic high microsatellite-instable (MSI-H) and hereditary non-polyposis coli cancer (HNPCC) colorectal cancer patients. Frequent MSI and heterogeneity in the MSI pattern were found both in normal and tumor cells from 10 HNPCC and sporadic MSI-H tumor patients whose tumors had defect mismatch repair protein expressions. This observation shows that MSI can also occur in nonneoplastic cells which has to be considered in MSI analyses for molecular HNPCC screening. In addition, considerable genetic heterogeneity was detected in all MSI-H (sporadic and HNPCC) tumors when analyzing five different regions with less than 150 cells, respectively. These differences were not detectable in larger tumor regions containing about 10,000 cells. Thus, heterogeneity of the MSI pattern (e.g. intratumoral MSI) is an important feature of tumors with the MSI-H phenotype.

Copyright © 2001 S. Karger AG, Basel

Introduction

Microsatellite instability (MSI) can be observed in a number of different types of tumors [1–6]. It is a specific feature for hereditary non-polyposis colorectal cancer (HNPCC). In order to define clinically HNPCC patients, the Amsterdam Criteria [7] have been developed. However, not all HNPCC patients with proven germline mutations of the mismatch repair genes hMSH2, hMLH1 and hMSH6 fulfill these criteria. For this reason, additional less stringent Bethesda guidelines [8] have been established. Patients fulfilling these criteria are potential candidates for HNPCC and should be tested for MSI in colorec-

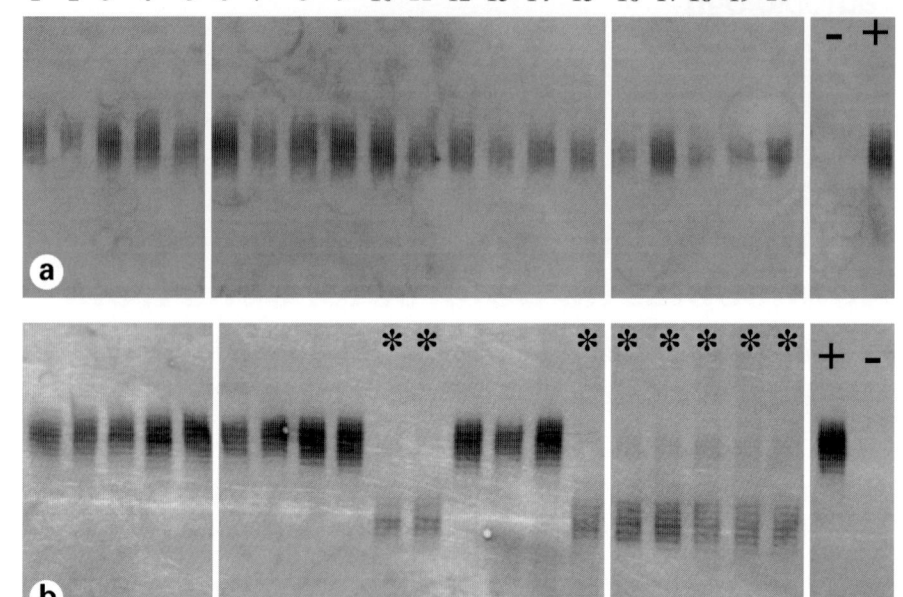

Fig. 1. Representative microsatellite analysis (BAT26) of laser-microdissected tissue samples containing 50–150 cells. PCR products were separated electrophoretically on 6.7% denaturating polyacrylamide gel. Lanes 1–5 represent normal cells, distance to tumor >5 cm, lanes 6–15 represent normal cells, distance to tumor <2 cm and lanes 16–20 represent tumor samples. + = Positive control; – = negative control. MSI is indicated by asterisks. Samples from MSS (**a**) and MSI-H (**b**) patients.

tal cancer (CRC). This test is done in parallel to an immunohistochemically performed mismatch repair protein expression analysis of hMSH2, hMLH1 and hMSH6. Because occurrence of MSI in colorectal tumors depends on the type of repeats within microsatellites, a sensitive and CRC-specific marker panel has been established [9], which is recommended by the 2nd HNPCC meeting as international reference panel [Bethesda reference marker panel, 10]. For each MSI analysis a matched pair of tumor and normal tissue sample is used, assuming that both tumor and normal cells are homogenous with respect to their microsatellite status. However, there are several reports which show that MSI is not restricted only to tumor cells. MSI has also been detected in nonneoplastic cells, e.g. in patients with chronic ulcerative colitis, hepatitis or liver cirrhosis [11–13]. Because the microsatellite pattern of normal cells is used as a reference for tumor MSI analysis, it is important to examine microsatellite alterations in normal colonic mucosa in HNPCC candidates. In this study, the potential heterogeneity in MSI was investigated by analyzing multiple samples of both normal and neoplastic cells from microsatellite-stable (MSS) sporadic and hereditary (HNPCC) high microsatellite-instable (MSI-H) colorectal tumor patients. Furthermore, MSI analysis was performed with small tissue samples (<150 cells) generated by laser microdissection, and compared to larger cell samples (>10,000 cells) in order to evaluate the influence of the small cell numbers on MSI results.

Material and Methods

Tissue Samples and Laser-Assisted Microdissection

Five MSS, 5 sporadic MSI-H and 5 HNPCC tumor patients were analyzed. For each patient at least 10 samples from normal colonic mucosa adjacent to the tumor, 5 tumor-distant normal samples as well as 5 different tumor regions were analyzed. 5-μm sections from formalin-fixed normal and tumor tissues were mounted onto a 1% poly-L-lysine-coated 1.35-μm polyethylene foil (PALM, Bernried, Germany) attached on a glass slide. The slices were subsequently deparaffinized and stained with hematoxylin-eosin. Microdissection of cell groups containing 50–150 cells was done by laser pressure catapulting (PALM, Bernried, Germany).

Microsatellite Analysis

Laser-microdissected tissue samples were used for whole genome preamplification by I-PEP as previously reported [14]. 2–5 μl of preamplified DNA was used for subsequent amplification of BAT26, BAT40, D5S346 and D2S123 microsatellite markers. PCR was done as described [9] with the exception of 50 amplification cycles instead of 35.

Results

MSI in Normal Colonic Epithelial Tissues

When normal cells from large areas (regions of at least 1 cm^2) were analyzed with the Bethesda reference marker panel, no MSI could be detected in samples of MSS or MSI-H tumor patients. In addition, no MSI was found in normal cells from MSS patients after laser microdissection of minute areas of normal mucosa containing less than 150 cells, regardless if cells adjacent to or distant

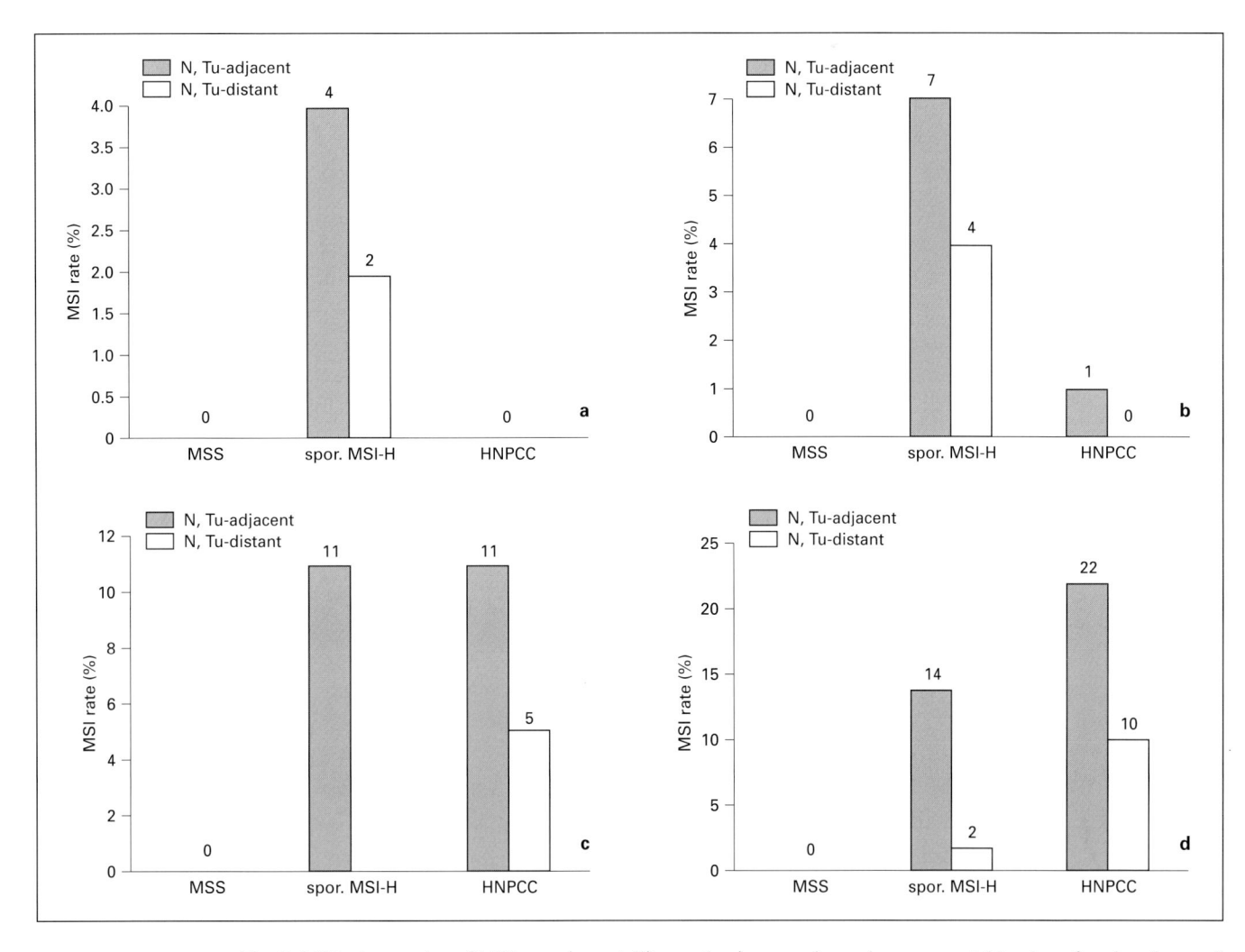

Fig. 2. MSI rate: number of MSI per microsatellite marker in normal samples as percent. Number of analyzed normal cells per lane is 50–150. N, Tu adjacent = Normal cells, distance to tumor <2 cm; N, Tu distant = normal cells, distance to tumor >5 cm; spor. = sporadic. **a** BAT 26. **b** BAT40. **c** D5S346. **d** D2S123.

from the tumor were investigated (fig. 1a). However, we could detect a considerable number of MSI in normal cells of MSI-H tumor patients using laser microdissection (fig. 1b). In general, frequency of MSI was higher in tumor-adjacent than in tumor-distant normal epithelial cells (fig. 2a–d). The highest MSI rates (number of MSI per microsatellite and group of tumors in percent) were found in regions in proximity to the tumor using the D2S123 marker in HNPCC patients (22%). The D5S346 marker exhibited also relatively high rates of MSI in normal cells (11% in both HNPCC and sporadic MSI-H CRCs). Interestingly, mononucleotide markers BAT26 and BAT40 showed a higher MSI rate in sporadic tumor patients than in HNPCC patients (4 and 7% vs. 0 and 1%, respectively).

Microsatellite Patterns in Tumor Tissues

Each investigated sporadic MSI-H cancer and four HNPCC tumors showed an MSI rate of 100% (MSI in 5/5 markers) when large tissue areas (regions of at least 1 cm^2) were analyzed using the Bethesda reference marker panel. Only one HNPCC tumor showed an 80% MSI rate (i.e. MSI in 4/5 marker). However, a notable heterogeneity in the microsatellite pattern could be observed in both hereditary and sporadic MSI-H tumors when clusters of 50–150 cells were analyzed. In general, heterogeneity was slightly, but not significantly higher in sporadic MSI-H than in HNPCC neoplasms. MSI rates as percents of MSI per microsatellite marker in 25 tumor samples in HNPCC and sporadic MSI-H tumors were 80 and 93% with BAT40, 76 and 84% with BAT26, 56 and 60% with

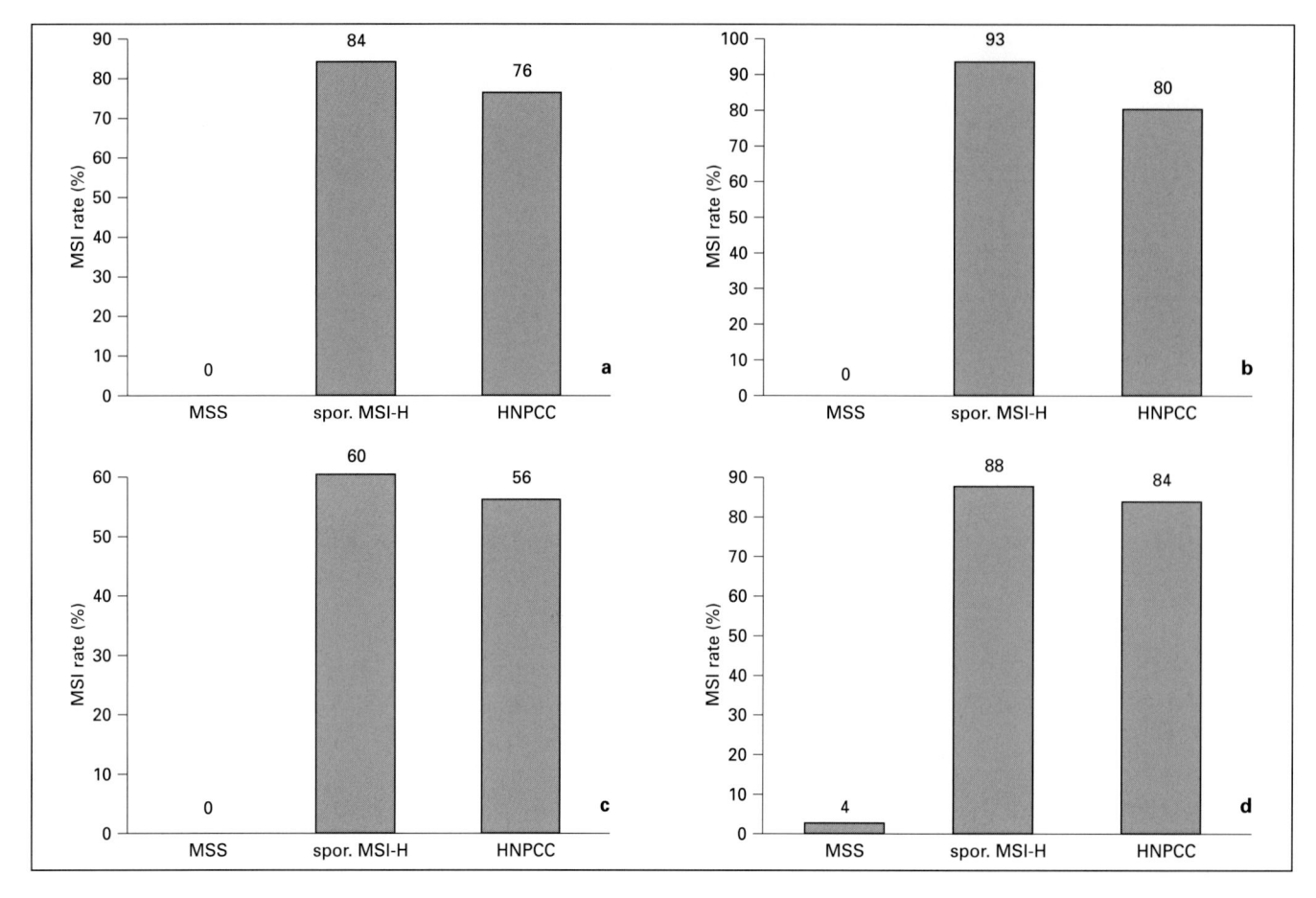

Fig. 3. MSI rate: number of MSI per microsatellite marker in tumor samples as percent. Number of analyzed cells per lane is 50–150. spor. = Sporadic. **a** BAT26. **b** BAT40. **c** D5S346. **d** D2S123.

D5S346, and 84 and 88% with D2S123, respectively (fig. 3a–d). A weak instability (4% with D2S123 marker) could even be detected in an MSS tumor.

Discussion

An important result of this study was a significant heterogeneity of the MSI rate in HNPCC and sporadic MSI-H tumor cell samples when a microsatellite analysis with small laser-microdissected cell groups (50–150 cells) was performed. This could not be observed in the same tumors if large cell samples (>10,000 cells) were used. In order to exclude that this heterogeneity was due to artificially generated spurious PCR bands the same analyses were done with MSS tumors as controls. In these samples, no MSI could be detected. Obviously, this finding demonstrates that MSI-H tumors contain a number of small subclones which do not all necessarily show microsatellite alterations. However, this phenomenon cannot be detected in larger tumor samples because each stable tumor region is masked by the occurrence of additional microsatellite bands after PCR and gel electrophoresis. For this reason an additive effect of MSI in different clones obviously takes place, which results in a higher MSI rate as compared to small samples which may consist of only a single clonal population. For molecular HNPCC diagnostics, the most sensitive strategy of microsatellite analysis is to examine larger carcinoma regions comprising many clonal populations rather than minute cell clusters. However, the tumor should be carefully microdissected to avoid a loss of sensitivity in detecting MSI due to contamination with normal tumor cells. Furthermore, intratumoral heterogeneity of MSI patterns is an important finding in tumors with the MSI-H phenotype.

In this study we investigated 5, 10 and 15 nonneoplastic cell samples of MSS, sporadic MSI-H and HNPCC colorectal tumor patients, respectively. Whereas no MSI could be detected in normal cells from MSS tumor patients, we found a considerable number of MSI in the normal mucosa in both sporadic and hereditary MSI-H tumor patients. Interestingly, MSI rate in normal cells was nearly twice as high in regions in close proximity to the tumor (<2 cm) as compared to regions distant from the tumor (fig. 2a–d). Our findings support the hypothesis that there are two molecular pathways in the development of colorectal tumors. In case of MSS tumors, instability of repetitive sequences does not play an important role. These tumors develop through a chromosomal instability pathway [15]. In contrast, tumorigenesis in both sporadic MSI-H and HNPCC tumor patients underlies the MSI pathway [15]. As a consequence tumor samples show a high frequency of MSI. However, the finding of MSI in normal colonic mucosa in MSI-H patients is an important result of our study and could argue for an early development of MSI in a field of morphologically normal-appearing cells. These may have undergone transient or manifested genomic damages or alterations as a consequence of factors like free radicals [16, 17] or inactivation of mismatch repair genes by promotor hypermethylation [18, 19], thus leading to MSI-positive cells. Some of these could aquire additional defects in genes regulating proliferation or apoptosis (e.g. TGF-β-RII [20] or BAX [21]) and eventually develop malignant tumor cells. These findings give new insights into very early steps of development of MSI-H colorectal tumors and have to be considered in molecular HNPCC diagnostics.

Acknowledgments

We thank Stephan Reinhold, Lisa Weber, Jutta Förster and Anja Trexler for excellent technical assistance.

References

1 Caduff RF, Johnston CM, Svoboda-Newman SM, Poy EL, Merajver SD, Frank TS: Clinical and pathological significance of microsatellite instability in sporadic endometrial carcinoma. Am J Pathol 1996;148:1671–1678.

2 Katabuchi H, van Rees B, Lambers AR, Ronnett BM, Blazes MS, Leach FS, Cho KR, Hedrick L: Mutations in DNA mismatch repair genes are not responsible for microsatellite instability in most sporadic endometrial carcinomas. Cancer Res 1995;55:5556–5560.

3 Horii A, Han HJ, Shimada M, Yanagisawa A, Kato Y, Ohta H, Yasui W, Tahara E, Nakamura Y: Frequent replication errors at microsatellite loci in tumors of patients with multiple primary cancers. Cancer Res 1994;54:3373–3375.

4 Toyama T, Iwase H, Yamashita H, Iwata H, Yamashita T, Ito K, Hara Y, Suchi M, Kato T, Nakamura T, Kobayashi S: Microsatellite instability in sporadic human breast cancers. Int J Cancer 1996;68:447–451.

5 Peng H, Chen G, Du M, Singh N, Isaacson PG, Pan L: Replication error phenotype and p53 gene mutation in lymphomas of mucosa-associated lymphoid tissue. Am J Pathol 1996;148:643–648.

6 Izumoto S, Arita N, Ohnishi T, Hiraga S, Taki T, Tomita N, Ohue M, Hayakawa T: Microsatellite instability and mutated type II transforming growth factor-beta receptor gene in gliomas. Cancer Lett 1997;112:251–256.

7 Vasen HF, Mecklin JP, Kahn PM, Lynch HT: The International Collaborative Group on Hereditary Non-Polyposis Colorectal Cancer (ICG-HNPCC). Dis Colon Rectum 1991;34:424–425.

8 Rodriguez-Bigas MA, Boland CR, Hamilton SR, Henson DE, Jass JR, Khan PM, Lynch H, Perucho M, Smyrk T, Sobin L, Srivastava S: A National Cancer Institute Workshop on Hereditary Nonpolyposis Colorectal Cancer Syndrome: Meeting highlights and Bethesda guidelines. J Natl Cancer Inst 1997;89:1758–1762.

9 Dietmaier W, Wallinger S, Bocker T, Kullmann F, Fishel R, Rüschoff J: Diagnostic microsatellite instability: Definition and correlation with mismatch repair protein expression. Cancer Res 1997;57:4749–4756.

10 Boland CR, Thibodeau SN, Hamilton SR, Sidransky D, Eshleman JR, Burt RW, Meltzer SJ, Rodriguez-Bigas MA, Fodde R, Ranzani GN, Srivastava S: A National Cancer Institute Workshop on Microsatellite Instability for cancer detection and familial predisposition: Development of international criteria for the determination of microsatellite instability in colorectal cancer. Cancer Res 1998;58:5248–5257.

11 Tsopanomichalou M, Kouroumalis E, Ergazaki M, Spandidos DA: Loss of heterozygosity and microsatellite instability in human non-neoplastic hepatic lesions. Liver 1999;19:305–311.

12 Park WS, Pham T, Wang C, Pack S, Mueller E, Mueller J, Vortmeyer A, Zhuang Z, Fogt F: Loss of heterozygosity and microsatellite instability in non-neoplastic mucosa from patients with chronic ulcerative colitis. Int J Mol Med 1998;2:221–224.

13 Cravo ML, Albuquerque CM, Salazar de Sousa L, Gloria LM, Chaves P, Dias Pereira A, Nobre Leitao C, Quina MG, Costa Mira F: Microsatellite instability in non-neoplastic mucosa of patients with ulcerative colitis: Effect of folate supplementation. Am J Gastroenterol 1998;93:2060–2064.

14 Dietmaier W, Hartmann A, Wallinger S, Heinmöller E, Kerner T, Endl E, Jauch KW, Hofstädter F, Rüschoff J: Multiple mutation analyses in single tumor cells with improved whole genome amplification. Am J Pathol 1999;154:83–95.

15 Lengauer C, Kinzler KW, Vogelstein B: Genetic instability in colorectal cancers. Nature 1997;386:623–627.

16 Richard SM, Bailliet G, Paez GL, Bianchi MS, Peltomaki P, Bianchi NO: Nuclear and mitochondrial genome instability in human breast cancer. Cancer Res 2000;60:4231–4237.

17 Jackson AL, Chen R, Loeb LA: Induction of microsatellite instability by oxidative DNA damage. Proc Natl Acad Sci USA 1998;95:12468–12473.

18 Kuismanen SA, Holmberg MT, Salovaara R, de la Chapelle A, Peltomaki P: Genetic and epigenetic modification of MLH1 accounts for a major share of microsatellite-unstable colorectal cancers. Am J Pathol 2000;156:1773–1779.

19 Kuismanen S, Holmberg MT, Salovaara R, Schweizer P, Aaltonen LA, de La Chapelle A, Nystrom-Lahti M, Peltomaki P: Epigenetic phenotypes distinguish microsatellite-stable and -unstable colorectal cancers. Proc Natl Acad Sci USA 1999;96:12661–12666.

20 Markowitz S, Wang J, Myeroff L, Parsons R, Sun L, Lutterbaugh J, Fan RS, Zborowska E, Kinzler KW, Vogelstein B, Brattain M, Willson J: Inactivation of the type II TGF-beta receptor in colon cancer cells with microsatellite instability. Science 1995;268:1336–1338.

21 Rampino N, Yamamoto H, Ionov Y, Li Y, Sawai H, Reed JC, Perucho M: Somatic frameshift mutations in the BAX gene in colon cancers of the microsatellite mutator phenotype. Science 1997;275:967–969.

Laser Microdissection as a New Approach to Prefertilization Genetic Diagnosis

Annette Clement-Sengewald[a] Tina Buchholz[b] Karin Schütze[c]

[a]I. Frauenklinik der Universität, and [b]Klinik und Poliklinik für Frauenheilkunde und Geburtshilfe, Klinikum Grosshadern, Ludwig Maximilians University, Munich, and [c]PALM Mikrolaser Technologie, Bernried, Germany

Key Words

Laser microdissection · Laser pressure catapulting · Laser tweezers · Laser trap · Polar body analysis · Prefertilization, genetic diagnosis · UV-A laser beam

Abstract

The genetic status of oocytes can be determined by polar body (PB) analysis. Following PB extraction, a genetic evaluation is performed. As each PB contains the complementary genetic material of the oocyte, PB analysis reveals information about its genetic status. Genetically altered oocytes may then be excluded from in vitro fertilization. The aim of our study was to evaluate laser microdissection as a tool for PB extraction purposes. Compared to the PB extraction with a sharp-ending pipette only, we could show that laser microdissection of the zona pellucida (laser zona drilling) with a UV-A laser and subsequent extraction with a blunt-ending pipette decreases the degeneration rate of oocytes. It is shown that laser pressure catapulting of extracted PB enables their contact-free transfer into tubes, thus decreasing the risk of contamination for further analysis.

Copyright © 2001 S. Karger AG, Basel

Introduction

Polar body (PB) analysis is a helpful procedure for the identification of oocytes bearing chromosomal disorders or mutations. As these altered oocytes will result in a failure of fertilization, abnormal embryo development, failure of implantation and early or late fetal loss, it would be necessary to identify these oocytes and exclude them from in vitro fertilization (IVF; prefertilization genetic diagnosis). As older women have a higher risk of chromosomal aberrations, a prefertilization genetic diagnosis may help to increase pregnancy- and baby-take-home rates especially in these patients. Furthermore, PB analysis may help patients with a known chromosomal disorder or mutation with otherwise minimal chances of a healthy baby. For PB analysis, PB have to be extracted from the perivitelline space of the oocyte. This has to be done by micromanipulation, either mechanically with micropipettes [1, 2] or with a combination of different methods, i.e. chemical treatment of the zona pellucida and subsequent mechanical withdrawal of the PB [3]. All methods bear the risk of damage to the PB or the oocyte. Therefore, optimization of the extraction method is one of the main tasks in prefertilization genetic diagnosis. As a degenerated PB may result in a false interpretation in subsequent genetic analysis [fluorescence in situ hybridization (FISH), single-cell polymerase chain reaction (PCR)], careful handling is a prerequisite for successful analysis. On the other hand, damage of the oocyte will alter the fertilization process and subsequent embryo development.

For a long time laser light has been used to manipulate cells or even cell fragments [4]. As laser beams may now be guided through optical lenses, it is possible to exactly focus the laser beam to less than 1 μm in diameter. In assisted reproductive technologies, laser light was used for manipulation of sperm and oocytes, thus supporting or replacing conventional manipulation methods [5–7].

First experiments in oocytes to microdissect the zona pellucida (laser zona drilling) were performed in order to facilitate sperm penetration [6, 8]. In the last years laser zona drilling has been reported to facilitate embryo hatching [9, 10]. In 1998, a diode laser was used to drill a hole into the zona pellucida of mouse oocytes and to subsequently extract PB by means of a blunt-ended pipette [11].

In this study we compared the conventional PB extraction method, using a sharp-ended pipette, with the combined extraction method of microdissection with a UV-A laser beam and PB extraction with a blunt-ended pipette. Furthermore, we tried to use laser tweezers for trapping and movement of PB to find out whether laser tweezers might be helpful for PB extraction.

In 1999, a new method for a noncontact, laser-mediated capture of single cells was published, the so-called laser pressure catapulting [12]. After isolation by laser microdissection, membrane-bound single cells were catapulted into PCR tubes. It was shown that this method did not impair subsequent DNA and RNA analysis.

For single-cell PCR contamination is very critical. Therefore, we tested the laser pressure catapulting method for the noncontact transfer of extracted PBs into PCR tubes.

Materials and Methods

Human Oocytes

For the experiments human oocytes were used which had failed to fertilize during the routine IVF program. Furthermore, surplus oocytes and oocytes showing abnormal signs of fertilization (one or three pronuclei) were used. Thus, oocytes were available 24 and 48 h after their retrieval. The experiments were accepted by the Ethics Committee of the Ludwig Maximilians University (No. 182/98). Patients who agreed with the use of their oocytes for the experiments gave us their written content.

Laser Equipment

For all experiments a PALM® Robot-CombiSystem was used (PALM Microlaser Technologies, Bernried, Germany), consisting of a pulsed, UV-A nitrogen laser (337 nm) for laser microdissection

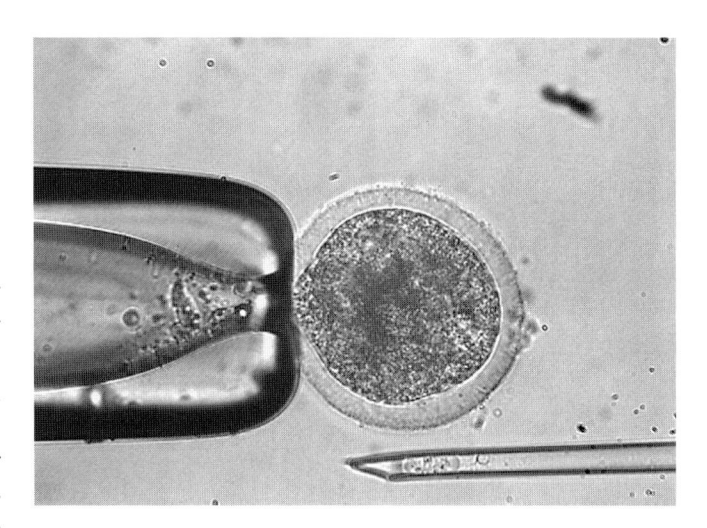

Fig. 1. PB extraction with a sharp-ended pipette (bovine oocyte). The oocyte was fixed with the PB in the 6 o'clock position. The pipette was inserted in the 5 o'clock position, and the PB was carefully sucked into the pipette. After withdrawal, the PB can be seen in the extraction pipette.

and a continuous-wave, diode-pumped Nd:YAG laser (1064 nm, 2000 mW) for laser tweezers purposes. The two laser beams were coupled into a microscope and their focus was adjusted to the optical focus of the microscope. All experiments were performed on a thin (0.17 mm) glass slide in a drop of 50 μl of manipulation medium (Ham's F10 medium +10% human albumin; ICN Biochemicals, Aurora, Ohio, USA). A 100-fold objective was used for all experiments.

Micromanipulation

Holding pipettes were pulled by hand, broken and subsequently fire-polished to an outer diameter of 150–180 μm and an inner diameter of 20–30 μm. For removal of the PB with a sharp-ended pipette, a borosilicate glass tube was pulled and broken in order to obtain an outer diameter of 15–20 μm at the tip. The tip was ground at an angle of 45°, and was subsequently washed in acid and water. Then, an additional peak was pulled to the tip with the help of a microforge. This peak guaranteed easy penetration of the zona pellucida. The blunt-ended pipette was pulled to an outer diameter of 15–20 μm, broken and fire-polished. For micromanipulation, the pipettes were connected to an air-filled (holding pipette) or paraffin oil-filled (extraction pipette) tube system. Their motion was controlled by mechanical micromanipulators.

For PB removal with the sharp-ended pipette, the oocyte was fixed by the holding pipette. The position of the PB was carefully considered to enable pipette penetration without injuring the oocyte. Therefore, the PB was in the 1 or 5 o'clock position (fig. 1), and the extraction pipette was inserted in the 2 or 4 o'clock position. The pipette was pushed between the PB and the oocyte, and the PB was carefully sucked into the pipette. After withdrawal of the pipette, the PB was released into a separate drop of medium. The oocyte was inspected 5 and 30 min after micromanipulation in order to evaluate its morphology.

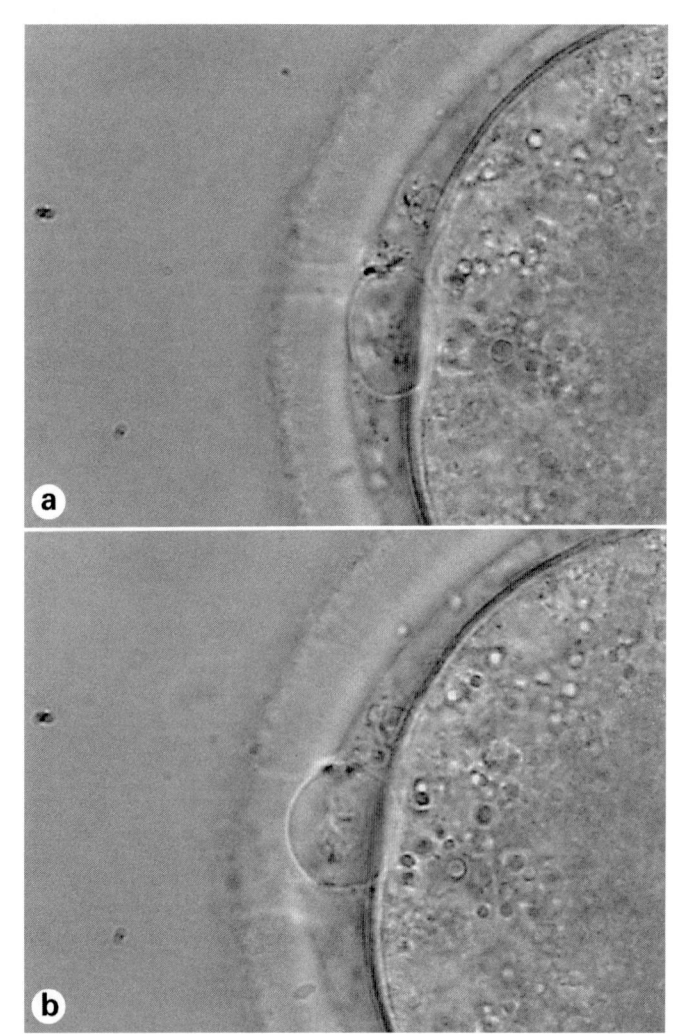

Fig. 2. Laser microdissection of the zona pellucida (laser zona drilling). **a** A small channel is drilled by moving the oocyte backwards and forewards into the UV-A microbeam. **b** After laser zona drilling, the hole was drilled slightly bigger than the size of the PB.

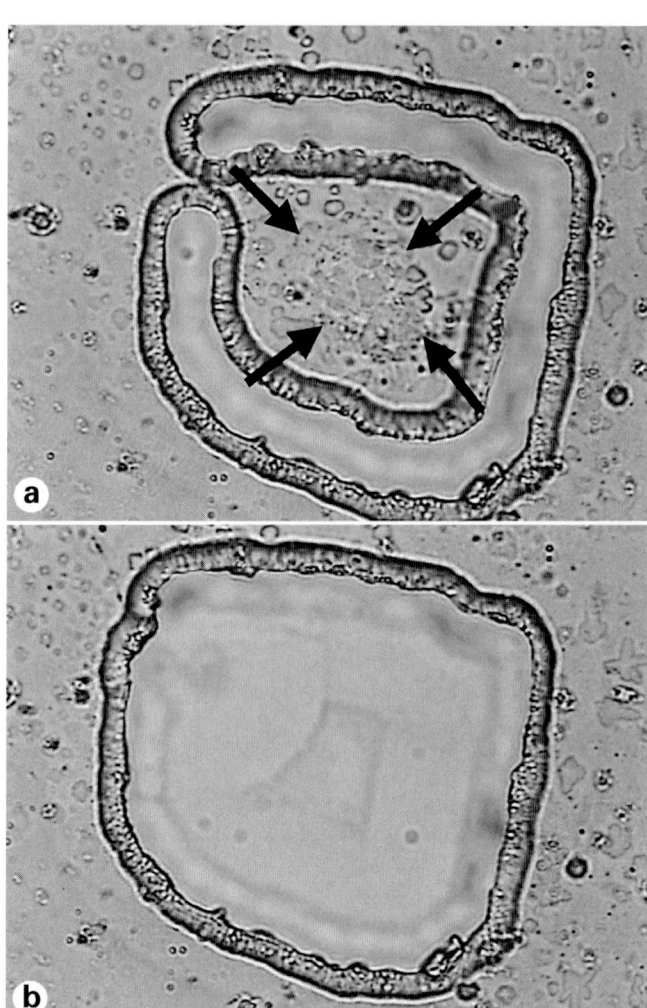

Fig. 4. Laser pressure catapulting of a human PB. **a** A UV-A laser beam cut the polyethylene membrane around the dried PB (arrows). **b** After laser pressure catapulting, with one single laser shot, the whole polyethylene membrane fragment was catapulted into the lid of the PCR tube.

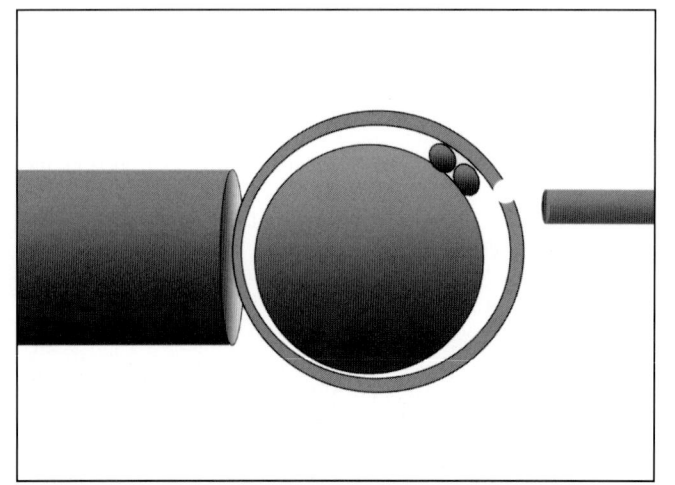

Fig. 3. Scheme of PB extraction by a blunt-ended pipette after laser microdissection of the zona pellucida. The extraction pipette can be easily inserted through the laser-drilled hole.

Table 1. Comparison of PB extraction with a sharp-ended and a blunt-ended pipette after laser microdissection of the zona pellucida (human)

Extraction method	Sharp-ended pipette only	Laser microdissection and blunt-ended pipette
Oocytes total	93	25
Oocytes with extracted first PB	46	18
Oocytes with extracted first and second PB	47	7
Intact oocytes 30 min after manipulation	67	24
Intact oocytes 30 min after manipulation, %	72	96

For PB removal with the blunt-ended pipette, the zona pellucida of the oocyte was first laser-microdissected with a UV-A laser beam. Therefore, the oocyte was set onto a glass slide in a small drop of manipulation medium. Close to the PB, a hole slightly bigger than the PB was drilled into the zona pellucida (fig. 2) by a high frequency of laser pulses. Therefore, the oocyte was moved forwards and backwards into the laser beam. As the focus of the beam was smaller than 1 µm, small paths in the zona were visible and allowed precise control of the manipulation procedure. When the hole was big enough, the blunt-ended pipette was introduced through the laser-microdissected hole (fig. 3) and pushed forwards to the PB. Then, the PB was carefully sucked into the pipette and subsequently released into a separate drop of manipulation medium. In a few oocytes we tried to catch single PBs using optical tweezers after zona laser microdissection. The trap was focussed on the PB to drag it out through the previously laser-drilled hole.

For laser pressure catapulting, an extracted PB was washed in a minimum amount of highly purified water and subsequently placed onto a thin polyethylene membrane, which had been mounted onto a glass slide (0.17 mm). During the drying process, the PB was constantly observed through the microscope to make sure that it could be identified again after drying. After complete drying, the polyethylene membrane around the PB was microdissected by a high frequency series of UV-A laser pulses (fig. 4), leaving a small bridge to avoid uncontrolled loss. With one single, high-energy pulse aimed to the joint, the membrane-bound PB was catapulted into the lid of a 0.5-ml plastic tube. For this purpose, the inner side of the lid had been covered with 1 µl of mineral oil. After catapulting, the membrane fragment stuck to the surface of the paraffin oil and its presence could be verified under a microscope.

Results

Microdissection of the Zona pellucida

Laser microdissection of the zona pellucida could easily be performed in all oocytes. In 25 human oocytes, no degeneration was observed after laser microdissection. This was confirmed in other experiments with bovine oocytes, in which 71 oocytes were laser-microdissected with no signs of damage or degeneration.

PB Extraction

Using a sharp-ended extraction pipette, the first PB was extracted in 46 oocytes and the first and second PB in 47 oocytes (table 1). Thirty minutes after manipulation, 67 of the 93 manipulated oocytes (72%) had remained morphologically intact. After laser microdissection of the zona pellucida in 25 human oocytes, the first PB was extracted in 18 oocytes and the first and second PB in 7 oocytes by a blunt-ended pipette. Twenty-four of the manipulated oocytes (96%) showed no signs of damage or degeneration 30 min after PB extraction. In other oocytes, it was possible to use optical tweezers to catch PBs and subsequently move them through the laser-drilled hole out of the perivitelline space.

Laser Pressure Catapulting

Laser pressure catapulting was applied in six human and fifteen bovine PBs. In all of them, laser microdissection of the polyethylene membrane around the PB was easy to perform and to control. If the laser focus was adjusted to less than 1 µm, the polyethylene membrane could exactly be microdissected around the PB without touching it. The energy for microdissection was between 1 and 3 µJ. Each PB could subsequently be catapulted into the lid of a PCR tube by a single UV-A shot using 10-fold the energy. Their presence on the surface of the paraffin oil was confirmed in all of the catapulted PB by control under the microscope.

Discussion

We have been able to show that the combined method of laser microdissection with a UV-A laser and PB extraction with a blunt-ended pipette decreases the degeneration rate of manipulated oocytes in comparison to PB extraction with a sharp-ended pipette only. Using this method, only 4% of the oocytes degenerated after manipulation, whereas 28% degenerated if a sharp-ended pipette was used for the PB extraction. In the literature, only results of an overall efficiency after genetic analysis of PB are given (FISH or PCR-related analysis), ranging between 60 and 82% [13–15]. Thus, it is impossible to compare our manipulation results with the results of other groups. In our experiments, FISH results of fixed PB can be interpreted in 67%. The overall efficiency, including oocyte degeneration rates, is therefore comparable with the published results of other groups.

The laser trap (laser tweezers) could be used to catch and move PB, and to extract them without pipettes from the oocyte. Although these were the very first trials with a new manipulation method for PB analysis, the first results are very promising. The main advantage of this method is the opportunity to work without any contact. This might dramatically decrease the risk of contamination, which is one of the most demanding problems in subsequent single-cell PCR and further analysis. Extracted PB, bound to a polyethylene membrane, could easily be microdissected and catapulted into PCR tubes. Thus, entire contact-free extraction and further procurement of PB seem to be probable in the future.

References

1. Verlinsky Y, Ginsberg N, Lifchez A, Valle J, Moise J, Strom C: Analysis of the first polar body: Preconception genetic diagnosis. Hum Reprod 1990;5:826–829.
2. Verlinsky Y, Cieslak J: Embryological and technical aspects of preimplantation genetic diagnosis; in Verlinsky Y, Kuliev AM (eds): Preimplantation Diagnosis of Genetic Diseases. New York, Wiley-Liss, 1993, pp 49–67.
3. Munné S, Scott R, Sable D, Cohen J: First pregnancies after preconception diagnosis of translocations of maternal origin. Fertil Steril 1998; 69:675–681.
4. Schütze K, Clement-Sengewald A: Catch and move – Cut and fuse. Nature 1994;368:667–669.
5. Tadir Y, Wright WH, Vafa O, Ord T, Asch R, Berns MW: Micromanipulation of sperm by a laser generated optical trap. Fertil Steril 1989; 52:870–873.
6. Clement-Sengewald A, Schütze K, Ashkin A, Palma GA, Kerlen G, Brem G: Fertilization of bovine oocytes induced solely with combined laser microbeam and optical tweezers. Hum Reprod 1996;13:259–265.
7. Germond M, Nocera D, Senn A: Improved fertilisation and implantation rates after non-touch zona pellucida microdrilling of mouse oocytes with a 1.48 μm diode laser beam. Hum Reprod 1996;11:1043–1048.
8. Liow SL, Bongso A, Ng SC: Fertilization, embryonic development and implantation of mouse oocytes with one or two laser-drilled holes in the zona, and inseminated at different sperm concentrations. Hum Reprod 1996;11: 1273–1280.
9. Antinori S, Selman HA, Caffa B, Panci C, Dani GL, Versaci CL: Zona opening of human embryos using a non-contact UV laser for assisted hatching in patients with poor prognosis of pregnancy. Hum Reprod 1996;11:2488–2492.
10. Germond M, Primi MP, Senn A, Pannatier A, Rink K, Delacrétaz G, Montag M, van der Ven H, Mandelbaum J, Veiga A, Barri P: Diode laser for assisted hatching: preliminary results of a multicentric prospective randomized study (abstract). Hum Reprod 1998;13:84–85.
11. Montag M, van der Ven K, Delacrétaz G, Rink K, van der Ven H: Laser-assisted microdissection of the zona pellucida facilitates polar body biopsy. Fertil Steril 1998;69:539–542.
12. Schütze K, Lahr G: Identification of expressed genes by laser-mediated manipulation of single cells. Nat Biotechnol 1999;16:737–742.
13. Rechitsky S, Strom C, Verlinsky O, Amet T, Ivakhnenko V, Kukharenko V, Kuliev A, Verlinsky Y: Accuracy of preimplantation diagnosis of single-gene disorders by polar body analysis of oocytes. J Assist Reprod Genet 1999;16: 169–175.
14. Verlinsky Y, Cieslak J, Ivakhnenko V, Lifchez A, Strom C, Kuliev A: Birth of healthy children after preimplantation diagnosis of common aneuploidies by polar body fluorescent in situ hybridization analysis. Fertil Steril 1996;66: 126–129.
15. Cieslak J, Ivakhnenko V, Evsikov S: Chromosomal abnormalities in first and second polar bodies (abstract). Meeting of the International Working Group on Preimplantation Diagnosis, Bologna, 2000, L-6.

Author Index Vol. 68, No. 4–5, 2000

Beyser, K. 227
Blümel, P. 218
Bock, O. 202
Bohle, R.M. 191
Borisch, B. 163
Buchholz, T. 232

Clement-Sengewald, A. 232

Dietmaier, W. 180, 227

Ermert, L. 191

Fend, F. 209
Fink, L. 191

Gänsbauer, S. 227
Glöckner, S. 173, 196, 202

Hartmann, A. 165, 180, 227
Hartmann, E. 191
Hiendlmeyer, E. 165
Hofstädter, F. 180, 227

Jauch, K.-W. 227

Kinfe, T. 191
Kleeberger, W. 173, 196
Knuechel, R. 165, 180
Kreipe, H. 173, 196, 202
Kremer, M. 209

Lahr, G. 218
Länger, F. 173
Lehmann, U. 173, 196, 202

Mürle, K. 165

Nathrath, W.B.J. 218
Nistér, M. 215

Olsson, Y. 215

Pösl, H. 218

Quintanilla-Martinez, L. 209

Ren, Z.-P. 215
Renke, B. 227
Rothämel, T. 196
Rümmele, P. 227
Rüschoff, J. 227

Sällström, J. 215
Schütze, K. 218, 232
Seeger, W. 191
Stich, M. 218
Stoehr, R. 165
Sundström, C. 215

Wieland, W. 165
Wild, P. 180
Wilke, N. 173

Subject Index Vol. 68, No. 4–5, 2000

Biopsies, archival 202
Breast cancer 173, 180

c-*erbB2* 173
Clonality 165
c-*myc* 173
Colorectal cancer 227
CyclinD1 173

DNA 215
Ductal carcinoma in situ 173

Endothelial cells 196
Endotoxin priming 191

Fluorescence in situ hybridization 165
Formalin-fixed material 215

Gene amplification 173, 202
Gene expression 209
Genetic diagnosis, prefertilization 232

Hepatocytes 196
Hereditary non-polyposis colorectal cancer 227
hMLH1 227
hMSH2 227

Immunohistochemistry 209, 227

Laser microdissection 173, 180, 196, 202, 218, 232
– pressure catapulting 218, 232
– trap 232
– tweezers 232

Laser-assisted cell picking 191
Lipopolysaccharide 191
Loss of heterozygosity 165
– – – analysis 180

Methylation 202
Microchimerism 196
Microdissection 165, 191, 215
Microsatellite instability 180, 227
Mismatch repair 227
mRNA quantitation 191

Nitric oxide synthase II mRNA 191

Papillary thyroid carcinoma 218
PCR 215
–, real time 173, 202
–, reverse transcription 202
– – –, real time 191
–, short tandem repeat 196
Polar body analysis 232
Polymerase chain reaction 209

RET/PTC1 218
RT-PCR 209

Tissue, archival 218
Topoisomerase IIα 173
Tumor heterogeneity 180

Urothelial lesions 165
UV-A laser beam 232

RB
43.8
.L3
L37
2001